Dornier Do 217

Dornier Do 217

From Bomber to Night Fighter: Rare Wartime Photographs

Chris Goss

First published in Great Britain in 2025 by
Air World
An imprint of Pen & Sword Books Limited
Yorkshire – Philadelphia

Copyright © Chris Goss 2025

ISBN 978 1 47388 309 3

The right of Chris Goss to be identified as
Author of this Work has been asserted by him in accordance
with the Copyright, Designs and Patents Act 1988.

A CIP catalogue record for this book is
available from the British Library.

All rights reserved. No part of this book may be reproduced, transmitted,
downloaded, decompiled or reverse engineered in any form or by any means,
electronic or mechanical including photocopying, recording or by any
information storage and retrieval system, without permission from the Publisher
in writing. No part of this book may be used or reproduced in any manner for
the purpose of training artificial intelligence technologies or systems.

Typeset by Mac Style
Printed in the UK by CPI Group (UK) Ltd, Croydon, CR0 4YY.

The Publisher's authorised representative in the EU for product
safety is Authorised Rep Compliance Ltd., Ground Floor,
71 Lower Baggot Street, Dublin D02 P593, Ireland.
www.arccompliance.com

For a complete list of Pen & Sword titles please contact

PEN & SWORD BOOKS LIMITED
47 Church Street, Barnsley, South Yorkshire, S70 2AS, England
E-mail: enquiries@pen-and-sword.co.uk
Website: www.pen-and-sword.co.uk
or
PEN AND SWORD BOOKS
1950 Lawrence Road, Havertown, PA 19083, USA
E-mail: uspen-and-sword@casematepublishers.com
Website: www.penandswordbooks.com

Contents

Acknowledgements vi
Introduction vii
Glossary and Abbreviations xv

Chapter 1   The Dornier Do 217: 1940–1941   1

Chapter 2   The Dornier Do 217: 1942   30

Chapter 3   The Dornier Do 217: 1943   114

Chapter 4   The Dornier Do 217: 1944–1945   192

Postscript   225

Acknowledgements

Yet again I would like to thank the continued help and support from my great friend Bernd Rauchbach. I have also had great help from Barbara, George and John Wolff (Canada and USA). I must also thank Günther Steudel (Canada) and the Steudel family. Sadly Günther passed away suddenly in 2022, and he has never seen how helpful he had been with all of my books on the Do 217.

In no particular order, I would also like to thank: Andy Saunders (UK), the late Dr Alfred Price (UK), Gianandrea Bussi (Italy), Mikael Olrog (Sweden), Peter Taghon (Belgium), Jeremy Green (USA), Del Davis (USA) and Ralf Lemser (Germany).

Chris Goss,
Marlow, 2025.

Introduction

Following the outbreak of war, Dornier's Do 17 'Flying Pencil' began to be replaced by more modern and capable bomber and reconnaissance aircraft. Indeed, it was as early as 1937 that Dornier began planning for the Do 17's replacement.

Design and Development

At the end of September 1938, the idea of a modernised Do 17 was first raised by the Luftwaffe. Full production of the Ju 88 had yet to be achieved, while much effort was now going into the Heinkel He 177 programme.

It was at the beginning of 1938 that Specification 1323 was presented. This called for a twin-engine bomber capable of long-range reconnaissance missions that would be powered by Daimler-Benz DB 601B engines. Dornier proposed a wider cockpit and a larger bomb bay capable of carrying 1,500kg of bombs and, compared to the Do 17 M, the wingspan was increased by a metre. The new aircraft was to be fitted with dive brakes.

At the same time, it was anticipated that this aircraft could also be used in the maritime attack role, as well as being able to be used as a heavy fighter. The Specification therefore had the aircraft fitted with floats, with an anticipated range of 1,500km and a solid nose with four fixed guns. However, arguments between the Luftwaffe and Kriegsmarine as to aircraft being land based or not, and emergence of other aircraft types, meant that development of a naval version, called the 'See-Stuka', soon ceased and the development of a more conventional aircraft continued.

The first prototype Do 217, V1, made its first flight on 4 October 1938. It crashed just one week later.

The first development aircraft, Do 217 V2, was ready to fly just under a month after V1 was lost. Between October 1938 and July 1940, ten prototypes flew, differing in the various designed roles (reconnaissance and bomber), and power plants. The final engines fitted, from the Do 217 V8 onwards, were BMW 801s.

At the same time, several other variants were being considered. Six Do 217s were produced, these being fitted with two jettisonable cameras and powered by DB 601B or F engines and with a crew of three. Likewise, nine Do 217 Cs were built powered by Jumo 211B or DB 601A engines and intended as bombers. However, both variants were secondary to what would become the first Do 217 to enter full operational service – the Do 217 E.

Into Service

Following the heavy losses suffered by the Do 17 units over Poland and then France, and, as it would transpire, the Battle of Britain, priority was given to bomber production and in particular the next generation of Dornier bombers. The Do 217 C became the basis for the Do 217 E-1, which, powered by BMW 801MA-1 engines, first flew on 1 October 1940, the same month that Do 17 production ceased. Surprisingly, the new design proved to be problem free.

By the end of March 1941, thirty-seven Do 217 E-1s had been built and test flown; a total of ninety-two would be completed before the arrival of the Do 217 E-2, of which another 185 would be built. The differences between both types were that the E-1 bomber was fitted with four MG 15 machine guns and a MG 151 for defence, while the E-2 dive bomber had three MG 15s and two MG 131s, one of which was fitted in a rear-firing dorsal turret, plus the fixed forward-firing MG 151.

Despite ongoing testing, which was not fully completed by the Luftwaffe until March 1942, the first unit to start to convert to the new type was II Gruppe/Kampfgeschwader 2 (II./KG 2). Most of the Gruppe returned to Achmer in Germany in March and April 1941. It was followed by III./KG 2 in September 1941 and I./KG 2 in November 1941.

Conversion complete, II./KG 2 moved to Evreux in France at the start of July 1941. It undertook its first mission on the night of 5-6 July 1941, the target on this occasion being RAF Pembrey in Wales. It was not long before the first loss was suffered. This occurred during a series of daylight attacks on shipping carried out on 14 July 1941. The aircraft in question, the Do 217 E-1 of 5./KG 2 flown by Feldwebel Kurt Bergmann, reported coming under attack by fighters near Land's End at 16.05 hours. It failed to return.

At around the same time, Flight Lieutenant Henryk Szczesny and Sergeant Stanislaw Brzeski of 317 Squadron reported shooting down a Ju 88 south of Tenby. The location, time and the fact that no other German aircraft were lost this day would indicate that the two Polish pilots had accounted for the first Do 217.

About now, a second Do 217 E-1 unit made an appearance. Formed in January 1941 as a maritime bomber unit flying the Heinkel He 111, in May 1941 II./KG 40 began converting to the Do 217. It would suffer its first combat loss on the night of 9 August 1941, when Oberleutnant Kurt Müller and his crew failed to return from a mission.

The RAF would have to wait until October 1941 to see the new type first hand. This occurred when Oberleutnant Günther Dolenga of 5./KG 2 put down his Do 217 E-1 near Rye in Kent on 12 October that year. Having been deceived by British beacons and, as a result, lost and running out of fuel, Dolenga took the decision to make a force-landing. The relative intactness of the aircraft gave the Royal Aircraft Establishment ample time to carry out a full technical assessment. It would be the first of many Do 217s to come down on British soil over the next two and a half years.

The Need for Change

Due to the increasing effectiveness of British defences, the Do 217 E needed to be improved. The E-3 (based on the E-1) and E-4 (based on the E-2) variants had additional armour protection for the crews and changes to defensive weapons and their calibre. The E-5 was based on the E-4, but was to be used for the Henschel Hs 293 glide bomb.

This was followed by the Do 217 K-1, essentially a Do 217 E-2, which, with a modified step-less cockpit and fitted with a nitrous oxide boost, had an increased range and speed. The Do 217 K-2, with its enlarged wing area, was an anti-shipping version capable of carrying two Hs 293 or PC1400 X (Fritz X) guided bombs, while the Do 217 K-3 was essentially the same as the K-2 but with modified bomb guidance equipment and bomb racks.

In time, however, shortages of BMW 801 engines forced another change. This resulted in the Do 217 M, which was fitted with the DB 603 in-line engine. Apart from the engines, it was almost identical to the K variant.

As well as the Do 217 E, both the K and M variants flew over the United Kingdom and the Mediterranean. Curiously, the only operational units to use the Do 217 on the Eastern Front were the Nachtaufklärungsstaffel night reconnaissance units – only occasionally did other units, such as the Einsatz Staffel/KG 101, an operational training unit, fly missions on the Eastern Front.

On the Western Front, II./KG 40 began converting to the Messerschmitt Me 410 in June 1943. This left just KG 2 to operate the Do 217; even then, by the spring of 1944, two of its three Gruppe were converting to the Ju 188. Similarly, I./KG 66, formed as a pathfinder unit in mid-1943, operated a limited number of Do 217 E, K and Ms, but by the spring of 1944, these had been replaced by aircraft such as the Ju 88 S.

One other unit which did well using the Do 217 was II and III./KG 100. Due to repeated delays into service of the He 177, which was to be a platform for launching Fritz X and Hs 293 guided missiles, it was decided that the Do 217 would help to bridge the gap. On 25 August 1943, twelve Do 217 E-5s from II./KG 100 took off from Cognac in France to attack an Allied convoy off the Spanish coast with Hs 293s. The attack resulted in damage to three warships. Two days later, another of the Do 217s sank the sloop *Egret*, while the destroyer HMCS *Athabaskan* was damaged.

Greater success came on 9 September 1943, when three crews from III./KG 100, operating from Marseilles-Istres, sank the Italian battleship *Roma* and severely damaged *Italia* west of Corsica.

In the aftermath of the Allied landings at Salerno, the American cruiser USS *Savannah* was sunk on 11 September 1943, and the cruiser USS *Philadelphia* damaged. Two days later, the cruiser HMS *Uganda* and destroyers HMS *Loyal* and HMS *Nubian* were similarly damaged, before, on 16 September 1943, the battleships HMS *Warspite* and HMS *Valiant* were also damaged.

There would be similar attacks off Anzio in January 1944, and both II and III./KG 100 would launch Fritz X and Hs 293 off the Normandy beaches in June 1944. However, despite the occasional success, II./KG 100 had lost in the region of twenty aircraft over Normandy by 5 July 1944. By the end of August that year, III./KG 100 had suffered similar losses, and both Gruppe withdrew to Germany and were disbanded.

Night Fighter

In response to the intensifying Allied bomber offensive, additional night fighters were increasingly needed by the Luftwaffe. The Do 217 E-2 was therefore modified by fitting four MG 17s and four MG-FF 20mm cannon in a solid nose. The rear-firing guns, including the MG 131 in the turret, were retained, as was the ability to carry bombs. This variant become the Do 217 J-1, which was intended as a night intruder. However, even before it first flew in October 1941, such missions over Britain ceased by order of the Führer. The J-1 was then fitted with FuG 202 Lichtenstein BC radar, the J-2 with FuG 212 Lichtenstein C-1.

Operational evaluation was carried out in March 1942 and, although found adequate, the Do 217 J-1 was only delivered piecemeal to various operational and training units. At the same time, some eleven of the type went to the Italian Air Force.

The J-1, though, was not popular, as one experienced night fighter pilot wrote: 'My Gruppe had a Staffel of Do 217s in 1943 because Bf 110s were in short supply and High Command thought that the four and half hour endurance compared to the two and a half of the Bf 110 could be of use. The 217 was fast, stable, excellent for instrument flying and obviously a very nice bomber but much too heavy on the controls for a fighter. I flew it once just to try it but after that I refused to use it on operations and stuck to my tried and tested 110 which was greatly superior as a fighter.'

In July 1942, the Do 217 N-1 first flew. This was identical to the J-2, but like the Do 217 M was powered by DB 603 engines. It would later have the turret and rear-facing guns removed, the N-2 then being fitted with obliquely-mounted upward-firing MG 151 cannon in the fuselage, the so called *Schräge Musik* modification.

By October 1943, at which point some 130 J-1 and J-2s had been built, as well as around 240 N-1 and N-2s (around ninety-five N-1s were converted to N-2s), production had ceased. All but a few operational units had handed over their Do 217s, though a number of training and Nachtaufklärungsstaffel, as well as elements of NJG 4 and NJG 100, continued to operate a number over the Western and Eastern fronts well into 1944.

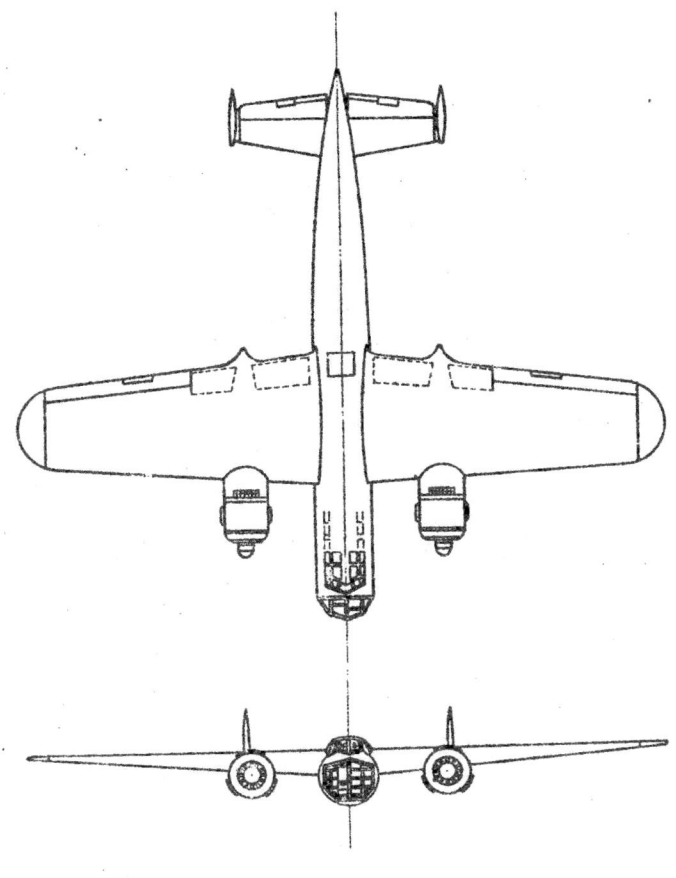

WEIGHT LB. (APPROX).	33,500
GROSS WING AREA. SQ.FT.	610
SPAN.	61·7

DORNIER. 217. E.1.

An early Air Ministry drawing of the Do 217 E-1.

xii Dornier Do 217

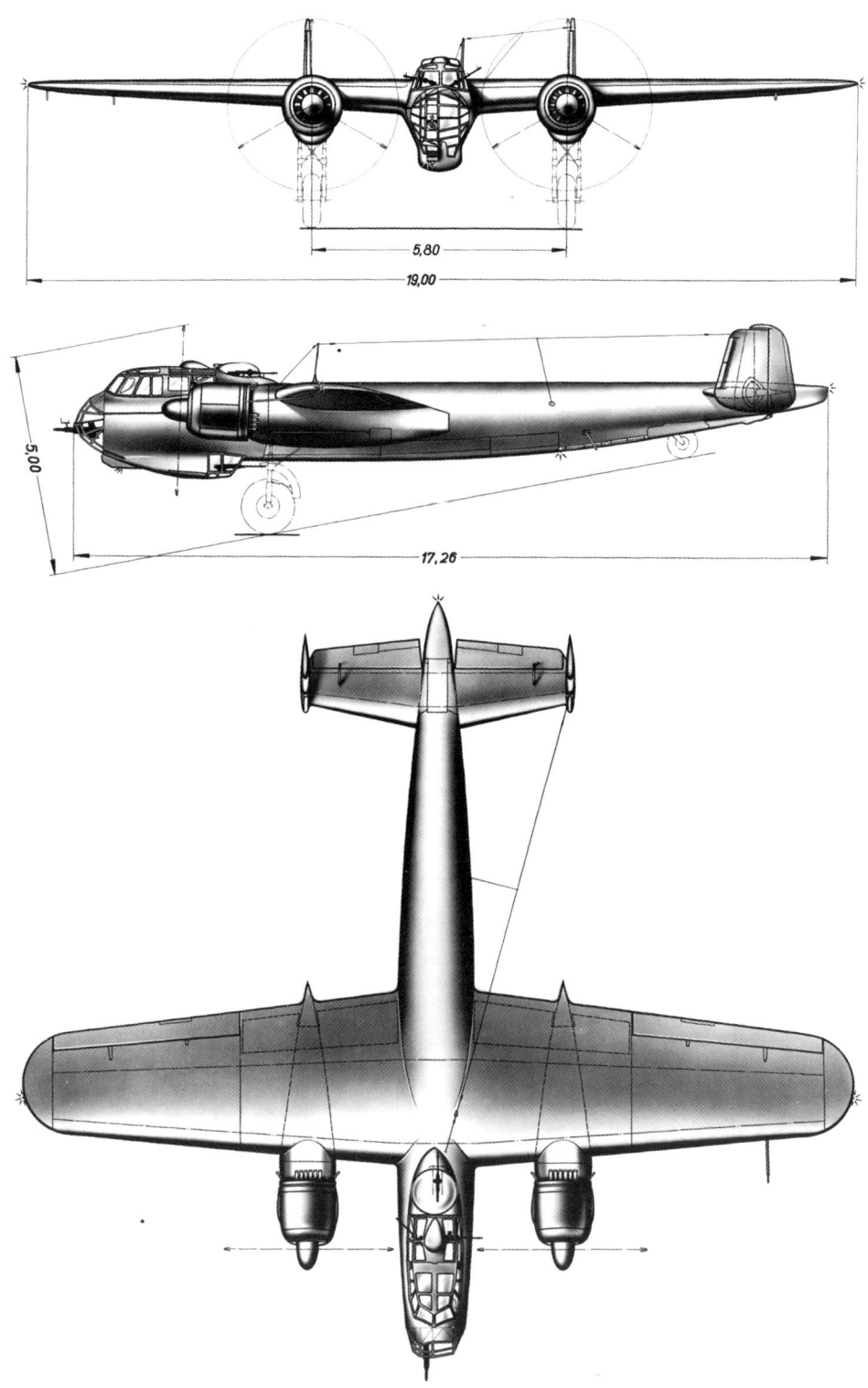

A later drawing of a Do 217 E-2/E-4.

A recognition guide for RAF pilots of the Do 217 E-2.

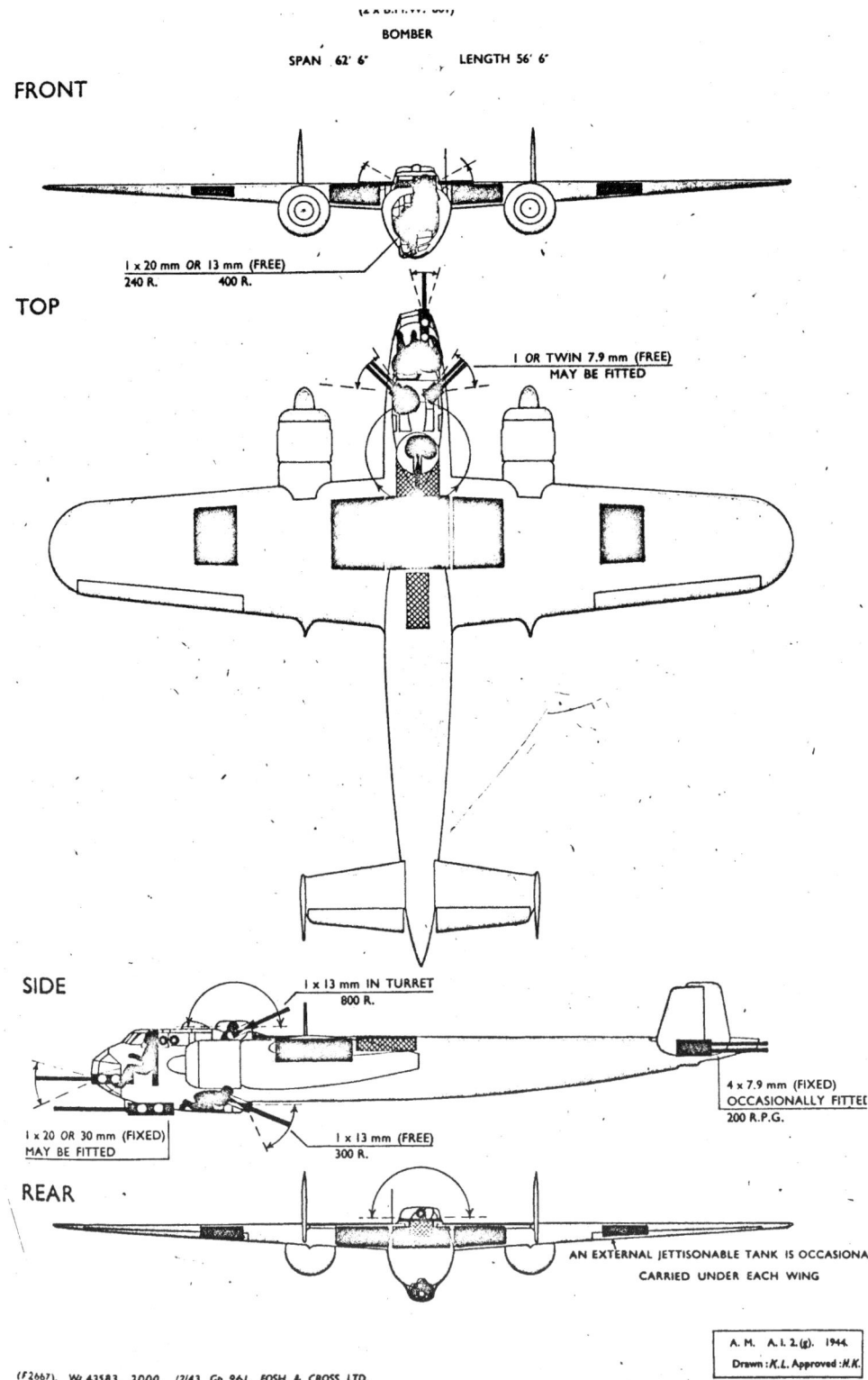

A drawing of the Do 217 E-4 dated 1944.

Glossary and Abbreviations

Aufklärungsgruppe	Reconnaissance wing
Bordfunker (BF)	Radio operator
Beobachter (BO or B)	Observer
Bordmechaniker (BM)	Flight engineer
Bordschütze (BS)	Air gunner
Deutsches Kreuz in Gold (DKiG)	German Cross in Gold
Ehrenpokal (Pokal)	Goblet of Honour, awarded for outstanding achievements in the air war
Eisernes Kreuz (EK)	Iron Cross (came in First and Second Class)
Ergänzungs (Erg)	Training
Erprobungs	Testing and development
Feindflug	Operational flight
Feldwebel (Fw)	Flight Sergeant
Fern	Long-range
Flak	Anti-aircraft
Flieger (Flg)	Aircraftman
Fliegerführer Atlantik	Air Commander for the Atlantic region
Flugzeugführer (F)	Pilot
Frontflugspange	Mission Class, awarded for different numbers of operational flights coming in bronze, silver and gold
Führer	Leader
Gefreiter (Gefr)	Leading Aircraftman
Generalfeldmarschall	Air Chief Marshal
Geschwader (Gesch)	Group consisting of three Gruppen, commanded by a Geschwader Kommodore (Gesch Komm)
Giesskanne	'Watering can', nickname for the tail mounted twin MG 81Z
Grossangriff	Massed attack
Gruppe (Gr)	Wing consisting three Staffeln; commanded by a Gruppen Kommandeur (Gr Kdr). Gruppe number denoted by Roman numerals
Hauptmann (Hptm)	Flight Lieutenant/Captain
IB	Incendiary bomb
Jagdgeschwader (JG)	Fighter Group
Kampfgeschwader (KG)	Bomber Group

Kette	Three aircraft tactical formation similar to RAF vic
Kommando (Kdo)	Detachment
Küstenfliegergruppe	Coastal Gruppe
Leutnant (Lt)	Pilot Officer/2nd Lieutenant
Luftflotte	Air Fleet
Luft Dienst (LD)	Communications and delivery unit
Luftmine	Air dropped mine
Lux	Air-dropped flame float which was ignited by chemical reaction between calcium phosphide and sea water
Major (Maj)	Squadron Leader
Nachtaufklärung	Night reconnaissance
Nachtjäger (NJ)	Night fighter
Nachrichenoffizier (NO)	Communications officer
Oberfeldwebel (Ofw)	Warrant Officer
Obergefreiter (Ogefr)	Senior Aircraftman/Corporal
Oberleutnant (Oblt)	Flying Officer/1st Lieutenant
Oberst	Group Captain/Colonel
Oberstleutnant (Obstlt)	Wing Commander/Lieutenant Colonel
Pirateneinsatz	Pirate attacks (single aircraft normally low level in daylight)
Reichsluftfahrtministerium (RLM)	Germany Air Ministry
Reichsmarschall	Marshal of the Air Force
Ritterkreuz (RK)	Knight's Cross
Ritterkreuz mit Eichenlaub (EL)	Knight's Cross with Oak Leaves
SC	Sprengbombe Cylindrisch, thin cased high explosive bomb
SD	Sprengbombe Dickwandig, thick walled fragmentation bomb
Stab	Staff or HQ; formation in which Gr Kdr and Gesch Komm flew.
Stabsfeldwebel (Stfw)	Senior Warrant Officer
Staffel (St)	Squadron (twelve aircraft); commanded by a Staffelkapitän Staffel number denoted by Arabic numerals
Störangriffe	Nuisance attack; sometimes called a Pirate attack
Technischer Offizier (TO)	Technical Officer
Trialen	German explosive similar to British torpex
Unteroffizier (Uffz)	Sergeant
Wettererkundungsstaffel (Wekusta)	Weather reconnaissance Staffel
Werk Nummer (Wk Nr)	Serial Number
Zerstörerangriff	Precision attack against a specified target

Chapter 1
The Dornier Do 217: 1940–1941

Dornier Do 217 V4, Wk Nr 690, the fourth prototype, pictured here at Friedrichshafen. It was also allotted the code CN+HL and was powered by Jumo 211 A-1 and then Jumo 211 B-1 engines. It is fitted with an MG 15 machine gun in the nose.

Dornier engineers working on a Do 217 E-2. The closest aircraft has the last two digits of a Wk Nr on the nose, which could indicate this was in fact Wk Nr 1152. This aircraft was damaged in an accident at Amsterdam-Schiphol while being flown by II./KG 2.

In the background can be seen Do 217 V8 Wk Nr 2708, coded CO+JL, which first flew on 21 March 1940. This was the test aircraft for the BMW 801 engines and was the first Do 217 E-1.

The Do 217 A-0, of which just six were produced, was fitted with cameras and assigned to 1./Aufklärungsgruppe Oberbefehlshaber der Luftwaffe from January 1940 for use in clandestine missions. They were powered by Daimler-Benz DB 601 engines and carried a crew of three.

Another view of the Do 217 A-0 seen in the previous image. Having been initially intended for clandestine missions, these aircraft were soon assigned to transport or training roles with other units. This aircraft is believed to be coloured pale grey and carries the T5 code on the fuselage.

The Dornier Do 217 V4. Behind it appears to be a dark-painted Do 217 E-1 or E-2, while in the foreground is the nose of either Do 17 V20, Wk Nr 2031, or V21, Wk Nr 2032.

Do 217 V7 Wk Nr 2707, variously coded D-ACBF and CO+JK, was originally fitted with BMW 139 engines but would later be refitted with BMW 801 engines and as such was the predecessor of the Do 217 E-1.

A Do 217 C-0, in this case that with the Wk Nr 2719. Only ten of this variant were built. This particular aircraft was powered by Junkers Jumo 211 engines. The last letters of the Stammkennzeichen are +TA.

Do 217 V9 Wk Nr 2709, seen here while fitted with BMW 801 engines.

A rear view of Do 217 E-1 Wk Nr 1006, coded DD+LF, showing the dive brake housing. This aircraft was damaged in a training accident on 11 April 1941. At the time, it was being operated by 6./KG 40 and coded F8+HP.

The first of two close-up view of a dive brake fitted on a Do 217 E-4. In this case it is pictured whilst closed.

This second picture of a dive brake fitted on a Do 217 E-4 shows it in the extended position.

The first of five official air-to-air press photographs of a Dornier Do 217 E-1 taken during a flight of snow-covered mountains.

The Dornier Do 217 E-1 is photographed here closing in on the camera ship. It is in fact the E-1 with the Wk Nr 1006 and which was coded DD+LF.

The Do 217 E-1 draws closer to the camera ship, in this case on the latter's starboard side. The Do 217 E-1 first flew on 1 October 1940.

The E-1's markings are clearly visible in this, the fourth photograph in the series. Some ninety-four examples of the E-1 variant were built.

At some point after this set of photographs was taken, the E-1 that formed its basis moved to 6./KG 40, where it became F8+HP.

The fully-loaded bomb bay a Do 217 E-1. In this case we can see two 500kg bombs in the foreground, with two 250kg bombs beyond. If required, the Do 217 E could carry four 500kg bombs internally.

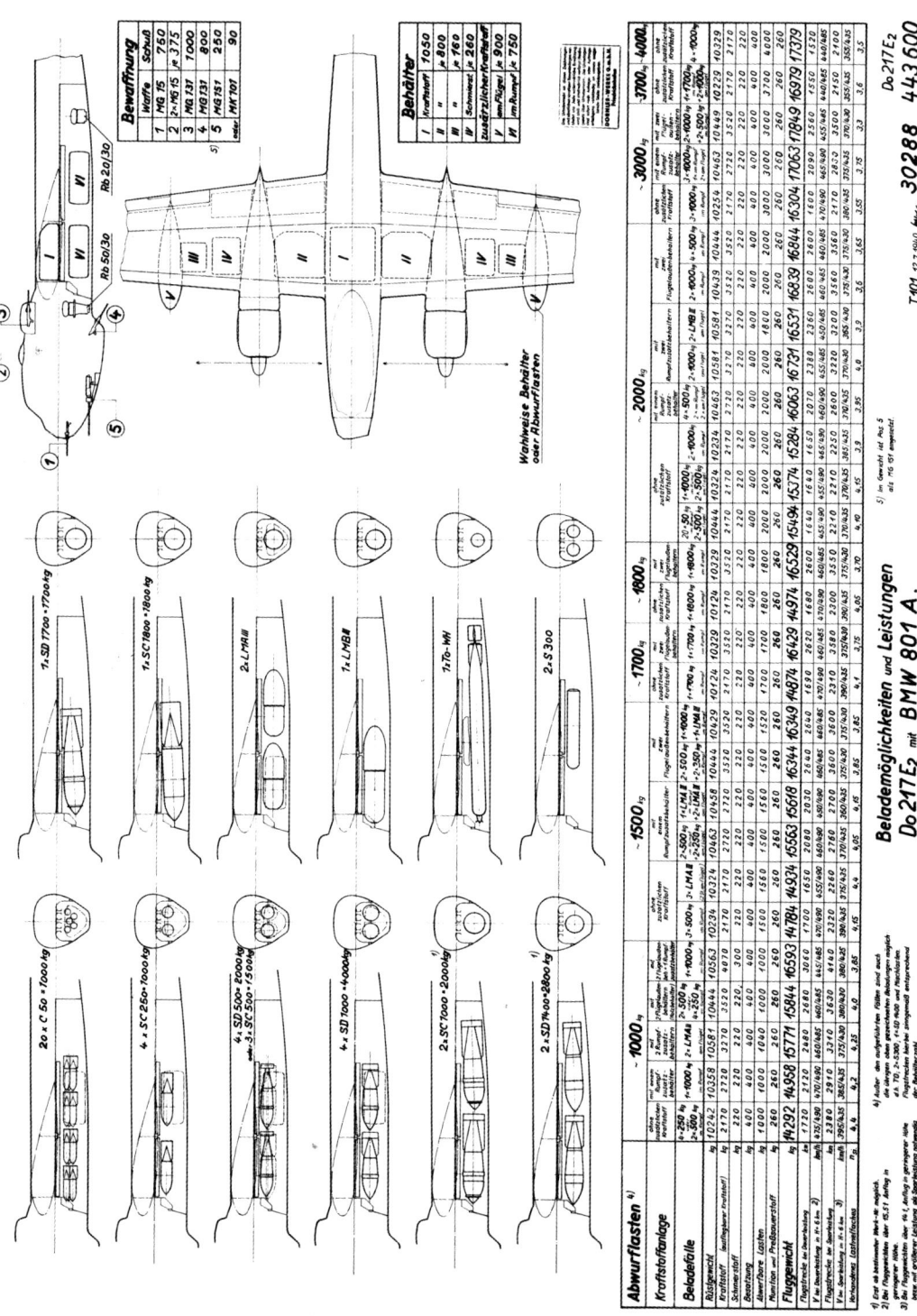

This Dornier document dated 12 July 1940 details the bomb load configurations and armament of the Do 217 E-2.

The first of five pictures that show Do 217 E-2 Wk Nr 1136, which was coded RH+EJ, during some flight trials.

In this second view of Do 217 E-2 Wk Nr 1136, note the turret type which differentiated the E-2 from the E-1.

This photograph of Do 217 E-2 Wk Nr 1136 clearly shows the elongated tail for the dive brake, which in this case is of course closed.

A line of Do 217 E-1s waiting to be delivered. Second from the right is Wk Nr 1020 which, coded as U5+BM of 4./KG 2, was written off in a crash-landing at Schiphol on 27 October 1941. Far right photo despite the censor's best efforts could either be Wk Nr 1033 or 1038 TC+ZL.

This aircraft is thought to be Do 217 E-2 Wk Nr 1145 U5+ZN of Hauptmann Heinz Engel's 5./KG 2. It has yet to have night camouflage applied so has possibly only just been delivered. Upgraded to an E-4, it was badly damaged when a tyre burst on landing at Vannes on 11 August 1943.

The Dornier Do 217: 1940–1941

New Do 217 Es awaiting delivery. At the rear on the left is believed to be Do 217 E-1 Wk Nr 1029 with the code letters TC+ZC – it was damaged in an accident at Evreux on 28 July 1941 while with II./KG 2. In the centre is an E-1, that with the Wk Nr 1008 and coded DD+LH. Lastly, at the front is believed to be Do 217 E-2 Wk Nr 1123 RE+CW, which was lost with 4./KG 2 as U5+EM when it was shot down by 307 Squadron on 1 November 1941.

A clear view of the Do 217 /E-4 turret and its 13mm gun.

The view looking rearwards out of a Do 217 E-4 dorsal turret.

Minor flak damage to the nose of a II./KG 2 Do 217 being inspected during 1941.

The first Dornier Do 217 to be captured almost intact was Oberleutnant Günther Dolenga's E-1, which had the Wk Nr 5069 and was coded U5+DN, of 5./KG 2. Confused by spoof radio beacons, Dolenga and his crew became lost during a mission on 12 October 1941. Thinking they were about 20 miles north-west of Le Havre, Dolenga opted to make a force-landing. This he did, only to discover that he had in fact come down on British soil – at Jury's Gap Sewer near Rye in East Sussex. The absence of a turret at the rear of the cockpit confirms that this was an E-1 variant.

The Dornier Do 217: 1940–1941   21

Oberleutnant Günther Dolenga of 6./KG 2 (second from left) and his crew, 1941.

Do 217 E-2 Wk Nr 1136 was later upgraded to an E-4 variant. It was lost in action while serving with 5./KG 2 on 4 February 1943. By then coded U5+DN, it crashed off Frinton-on-Sea in Essex, killing Feldwebel Hans Menzel and his crew.

The Dornier Do 217: 1940–1941

The first of four photographs taken of Dornier Do 217 E-2, Wk Nr 1144, which was coded RH+ER.

A starboard view of Do 217 E-2, Wk Nr 1144. Note the '44' marked in the centre of the *Balkenkreuz*.

Do 217 E-2 Wk Nr 1144 was lost in action on 22 March 1943. Coded U5+IN and being operated by 5./KG 2 it was shot down during an attack on Hartlepool.

Our last picture of Do 217 E-2 Wk Nr 1144. It is believed that the victor on 22 March 1943, was Flying Officer Robert Sargent of 219 Squadron. The Dornier is thought to have come down in the sea off Cullercoats, killing Feldwebel Rudolf Wenkel and his crew.

Do 217 E-2 Wk Nr 1144 was lost in action on 22 March 1943. Coded U5+IN and being operated by 5./KG 2 it was shot down during an attack on Hartlepool.

Our last picture of Do 217 E-2 Wk Nr 1144. It is believed that the victor on 22 March 1943, was Flying Officer Robert Sargent of 219 Squadron. The Dornier is thought to have come down in the sea off Cullercoats, killing Feldwebel Rudolf Wenkel and his crew.

Do 217 E-2 Wk Nr 1145, RH+ES. Again, note the last two digits of the Wk Nr inside the fuselage cross. This aircraft was later upgraded to a Do 217 E-4 and was badly damaged at Vannes on 11 August 1943 when a tyre burst on landing. In use with 5./KG 2, by that time it had been coded U5+ZN. It was being flown by Unteroffizier Ulrich Bachmann who, along with the rest of his crew, was uninjured.

An atmospheric night-time shot of Do 217 E-2 Wk Nr 1138, coded RH+EL. It has the last two digits of the Wk Nr on both the fuselage cross and nose. It, too, was upgraded to a E-4 variant. It was damaged in an accident at Salon, France, on 18 April 1943. It was being flown by II./KG 101 at the time.

A Dornier Do 217 E-4, that coded GG+PA, pictured while undergoing trials. Note the last two digits of the Wk Nr, in case '71', on the nose.

A Do 217 E probably awaiting delivery to KG 2 or II./KG 40. This might be the E-1 which, with the Wk Nr 1044, was coded TC+ZR.

A Do 217 E-2, that coded F8+IM and with the Wk Nr 1126, of 4./KG 40. This aircraft force-landed at Quakenbrück, having suffering 50 per cent combat damage, on 27 October 1941. Repaired and converted to an E-4, it returned to service before then being recorded as suffering an engine failure, with 11./KG 2, on 22 January 1943.

Chapter 2
The Dornier Do 217: 1942

A pair of heavily toned down Do 217 Es, presumably from KG 2, pictured at Gilze-Rijen.

A Do 217 E-4 of III./KG 2 photographed in a hangar at Gilze-Rijen during 1942.

Two Do 217s of 7./KG 2 approaching the Eiffel Tower while sightseeing Paris from the air, 1942.

Oberleutnant Rolf Häusner of 7./KG 2 speaking to air and groundcrew while behind is Do 217 E-4, which was coded U5+BR.

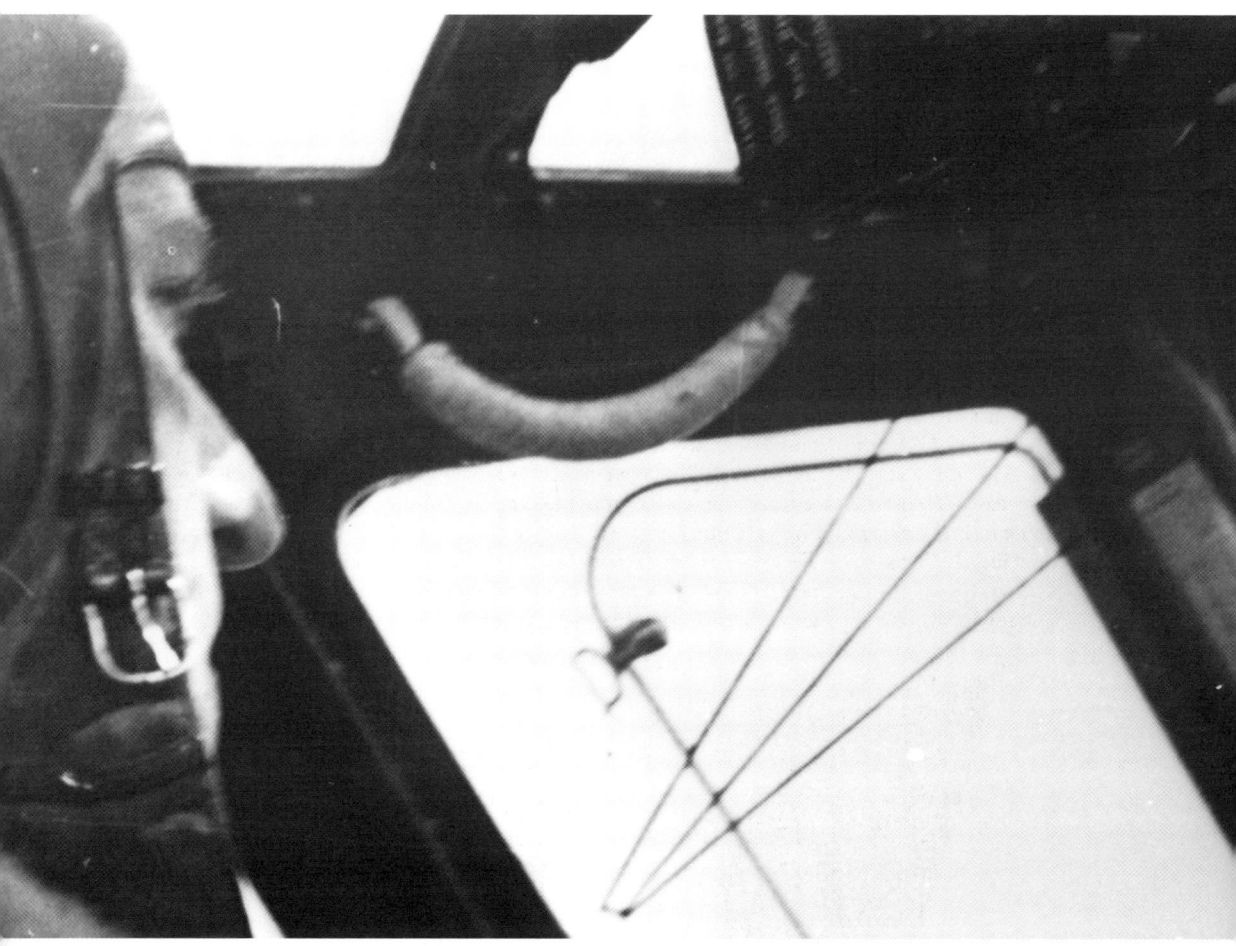

Oberleutnant Rolf Häusner photographed at the controls of his Do 217 E-4. Note the lines on the cockpit window, which were used for ascertaining diving angles.

This time it is just Oberleutnant Rolf Häusner's right hand which appears in this picture taken while he was flying his Do 217 E-4. One dramatic attack carried out by Häusner saw him target a train approaching Bramley station in Surrey at 14.10 hours on 16 December 1942. A young girl on board the train later recalled the events that day: 'When we slowed down on our approach to Bramley Station, I suddenly heard machine guns hitting the train. Looking up I had a fleeting glance of a plane out of the window headed for Guildford. It was very low just above the trees. Then there was a big explosion and the train rocked on the rails. The door on the far side was blown out and the glass flew across the compartment… Everything went completely quiet and still. I got my hands over my face before the glass blew in and had no bad damage to my face, although my hands were cut. Many people were not so lucky, several lost eyes.'

This picture of Luftwaffe ground crew at work on U5+BR in a hangar at Gilze-Rijen provides a close-up of the cockpit area of a Do 217 E-4.

Oberleutnant Rolf Häusner's Do 217 E-4, U5+BR, of 7./KG 2. This is another photograph taken at Gilze-Rijen in 1942. Häusner was awarded the Ehrenpokal on 16 November 1942, but was killed one month and a day later, on 17 December 1942 – the very day after his attack at Bramley. Häusner and his crew met their ends when they flew into a hillside as they were bombing York. Startlingly, in the wreckage, RAF investigators found a 1:500,000 map of Southern England, on which a rectangle marked precisely the location of Bramley.

Members of the crew of a 7./KG 2 Do 217 E-4 about to board their aircraft for a mission in 1942.

An unidentified Do 217 E-4 photographed in 1942.

An air-to-air shot of Do 217 E-4 Wk Nr 5342. Flown by 7./KG 2, this aircraft was coded U5+KR. It was probably shot down by Flying Officer Yvan du Monceau of 609 (West Riding) Squadron on 18 February 1942.

Do 217 E-4s of III./KG 2 carrying out formation practice. Note how the Holzhammer and diagonal band on the noses have been toned down.

Unlike in the previous image, the band and Holzhammer badge are clearly visible on this Do 217 E-4 of III./KG 2 that was photographed at Gilze-Rijen in 1942.

A Do 217 E-4 running up its engines at Gilze-Rijen in 1942. Note the heavily-camouflaged hangar in the background.

Luftwaffe personnel inspecting, or working on, a Do 217 E-4 at Gilze-Rijen, once again in 1942.

A Do 217 E-4 of II./KG 40 with a deadly cargo, in the form of 250kg bombs, waiting in the foreground.

46 Dornier Do 217

Bombs being delivered to III./KG 2 at Gilze-Rijen, 1942.

This Do 217 E-4 is believed to be from 8./KG 2 and coded U5+GS. The haphazard toning down and numeral on the tail is a puzzle.

The Dornier Do 217: 1942

Unteroffizier Horst Arnscheidt of 7./KG 2 in the cockpit of Do 217 E-4, Wk Nr 5458 which was coded U5+GR.

Another view of of Do 217 E-4 Wk Nr 5458. This aircraft was lost in action on 29 May 1942. At the time it was being flown by Feldwebel Heinz Jaros.

A Do 217 E-4 of I./KG 2 carrying out training over the Bay of Biscay, June 1942.

The wreckage of Do 217 E-4 Wk Nr 5372, coded U5+DL, which crashed at La Gorgue, a couple of miles east of Merville, in the early hours of 16 June 1942. At the time it was returning from a minelaying sortie off the Isle of Wight. Oberleutnant Klaus-Jürgen Schmitz, who had been awarded the Ehrenpokal only the previous day, and all of his crew were killed. Note the letters DL upside down on what was the tail.

An accident at Cognac in 1942. It is possible that both aircraft come from IV./KG 2, as each carries a letter on the tail which was particular to training units.

Dornier Do 217 aircrew chatting at Cognac in the spring of 1942. Left to right are Hauptmann Mitscher (Stab/KG 2), Leutnant Hans Peters (4./KG 2; killed on 28 July 1944), Leutnant Anton Herrmann (6./KG 2), and Oberleutnant Friedrich Dörflinger (6./KG 2; killed on 30 July 1942).

The Dornier Do 217: 1942   51

As mentioned in the previous caption, Oberleutnant Friedrich Wilhelm Dörflinger was killed on 30 July 1942, when his Do 217, that coded U5+DP, was shot down by a Beaufighter from 68 Squadron during an attack on Birmingham. The latter was piloted by Flying Officer Eric Raybould. The bomber's wreckage, seen here, came down at Salthouse Marshes, Norfolk.

Another view of the wreckage of Dörflinger's Do 217 E-4, which had the Wk Nr 1213, at Salthouse Marshes, Norfolk. Dörflinger had been awarded the Deutsches Kreuz in Gold on 18 May 1942.

The first of three pictures of a Do 217 E-4 of 5./KG 40, in this case that with the Wk Nr 4279 and coded F8+CN.

The starboard side of Do 217 E-4 Wk Nr 4279. Based at Soesterberg, in the Netherlands, this aircraft was shot down during a night attack on an aircraft factory in Bedford on the night of 23 July 1942. Oberleutnant Heinrich Viess and his crew all baled out to become prisoners of war. Their aircraft crashed near Spalding.

A stunning, and somewhat unusual, shot of the nose of a Do 217 E-4, this being the last of our pictures relating to F8+CN. The loss of this aircraft on 23 July 1942, was jointly claimed by Flight Lieutenant Harvey Sweetman, at the controls of a Hurricane of 486 Squadron, and Flight Lieutenant Ted McMillan, who was flying a Beaufighter from 409 Squadron.

The remains of the dorsal turret of a Dornier Do 217 E4, Wk Nr 5482 and coded U5+FP, that was shot down by anti-aircraft fire in the early hours of 31 July 1942. Flown by Hauptmann Alfred Kindler, the aircraft crashed at Newmarket; Kindler and his crew all baled out and were captured. (via Andy Saunders)

A close-up of the dorsal turret remains from U5+FP. Kindler's aircraft was one of three Do 217s lost that night, all three being from 6.Staffel based at Eindhoven. The raiders' target had been Birmingham. For his part, Kindler had survived some 230 combat sorties. (via Andy Saunders)

One Do 217 E-4 pilot, Oberleutnant Wilhelm Schmitter, seen here on the left, pictured with his crew in front of their aircraft. (via Jeremy Green)

Oberleutnant Wilhelm Schmitter posing for the camera in front of his Do 217 E-4, which was coded F8+BC. It is sporting a type of pale camouflage scheme that was worn by aircraft regularly used in harassing attacks. Officially called 'Störangriff', such missions were more commonly referred to as 'Pirateneinsatz' (pirate operations). They were best described by a German airman who was captured in August 1942, who then revealed the following: 'Suitable targets are selected many weeks in advance and intensive preparations are made. Very careful thought is given to the choice of route, large-scale maps and any available photographs of the target, and the approaches to it are closely studied – III./KG 2 even had a special sand table prepared for use in connection with pirate raids. Pains are taken to work out a route that runs over flat country and avoids any hills or natural or artificial obstructions. Towns are studiously avoided. A hand-drawn sketch is frequently made covering a strip eight miles wide on either side of the route, and all of the crew are expected to memorise the landmarks along it. As many as three aircraft may be despatched to attack an important objective, but it is considered preferable to decide upon several different targets to be attacked by individual aircraft on the same day.' (via Jeremy Green)

The tail of Schmitter's F8+BC showing silhouettes denoting two attacks, both of which had been on the town of Leamington Spa. It was during one of these, on 16 July 1942, that Schmitter was wounded by splinters from anti-aircraft fire; he was placed on leave for a month to recuperate. (via Jeremy Green)

Oberleutnant Schmitter's F8+BC running up its engines prior to another sortie over the UK. Credited with at least 4 victories and 193 missions in the west, many over 4 hours in duration, Schmitter was killed when the Messerschmitt Me 410 he was flying was shot down and crashed near Eastbourne in November 1943. (via Jeremy Green)

A technical problem caused this Do 217 E-4 Wk Nr 4313, of the Luftflotte Reserve for Luftflotte 3, to force-land in a field at Beez on 7 August 1942.

This is believed to be Do 217 J-1 Wk Nr 1290, which was coded D5+OM, of 4./NJG 3. This aircraft crashed into the sea just off Wenningstedt on 9 August 1942. Leutnant Günther Martens and his two crew were killed.

This unidentified Do 217 E-4 has suffered extensive damage during a crash landing. Note the letter C on the wing leading edge.

A Do 217 E-4, that coded F8+DM, of 4./KG 40 is pictured during a low-level flight. Two aircraft with this code were lost in 1942; Wk Nr 5422 on 10 August 1942, and Wk Nr 5537 some four months later on 22 December 1942.

Coded F8+AP, this Do 217 E-4 was pictured while being operated by 6./KG 40. Although it is on the tail, the Wk Nr cannot be discerned.

A second view of a Do 217 E-4 coded F8+AP. A total of four aircraft were recorded as carrying this code. The first is Wk Nr 5368, which was lost on 22 April 1942. Wk Nr 5502, meanwhile, was lost on 8 September 1942, and Wk Nr 5542 on 2 October 1942. The last was Wk Nr 4384, which met its fate on 18 December 1942.

Reasonably extensive combat damage to the tail of Do 217 E-4 F8+EM, of 4./KG 40, in August 1942.

Aircraft of Lehr und Erprobungskommando 17 at Chartres, France, in September 1942.

A Do 217 E-4 of Erprobungskommando XY in a hangar at Chartres in 1941 or 1942.

Personnel and aircraft of I./KG 66 pictured at Chartres whilst the unit was training. On the right is Leutnant Hans Altrogge.

Sergeant Ed Fox of 406 Squadron poses with the remains of the tail of the Do 217 E-4 of 1./KG 2 which crashed at St Just, Cornwall, on 26 September 1942. The Dornier had been tasked to attack a factory in Penzance when it was attacked and shot down by Squadron Leader Dennis Furse and Pilot Officer J.H. Downes in a Beaufighter of 406 Squadron. With Furse unavailable, Fox, who claimed a Ju 88 probably destroyed that night, was asked to stand in for Press. (via Phil Irwin)

This Do 217 E-4 of 8./KG 2, Wk Nr 4289 and coded U5+LS, crashed on Southend airfield on 26 October 1942. One account states that the Dornier was tasked with attacking the Skefko ball-bearing works at Luton. But having dropped four bombs near Rochford Hospital, it then turned over the aerodrome, at which point it was engaged by light anti-aircraft fire. Hit and out of control, the Dornier slammed into the ground and skidded across the surface to crash into a dispersal hut. The pilot, Unteroffizier Rudolf Schumann, was killed or wounded, along with two of his crew. The fourth member was flung clear, but died in hospital on 25 January 1943. One man on the ground, Warrant Officer Theodulle Dyon of 350 (Belgian) Squadron, was killed, two other groundcrew wounded.

The tail band on this Do 217 E-4 of 3./KG 2 pictured in November 1942, that coded U5+FL, denotes that it was assigned to Operation *Stockdorf*. The latter was the Luftwaffe's response to Operation *Torch*, the Allied invasion of French North Africa. Under *Stockdorf*, Luftwaffe bomber, reconnaissance and fighter units were hastily relocated to southern France. With the Wk Nr 4377, U5+FL was lost during an attack on London on the night of 8 February 1943. Due to the badge on its nose, the aircraft in the background is believed to be U5+GS of 8./KG 2.

Another Do 217 E-4, in this case Wk Nr 5562 which was coded U5+LS, clearly showing the distinctive Operation *Stockdorf* recognition markings. This aircraft suffered an engine failure shortly after taking off to attack Hull on 9 March 1943, and was destroyed in the subsequent crash-landing at Alsmeer. Feldwebel Wilhelm Haase and his crew were all injured.

As well as showing a presence during Operation *Stockdorf*, hence the medium to low-level flying, Luftwaffe aircraft such as these Do 217s were intended to help reinforce German efforts to prevent the scuttling of the French Fleet in Toulon.

Do 217 E-4s of 9./KG 2 taxiing out. The aircraft second from the left is that coded U5+LT, while the next aircraft in the line is U5+FT; both have the last two letters on their tails. They also have a fuselage band, but there is no sign of *Stockdorf* engine and wing markings.

Two Do 217 E-4s of 8./KG 2 flying over the Pyrenees in November 1942. Both are displaying the Operation *Stockdorf* markings.

Almost certainly photographed in the same area and in the same flight as the previous picture, this photograph shows the same pair of Do 217 E-4s of 8./KG 2.

In this image of two Do 217 E-4s pictured at their base, the furthest aircraft is that coded U5+KS. It is still in Operation *Stockdorf* markings. The nearest aircraft carries the 8.Staffel pennant usually denoting the Staffelkapitän's aircraft, who at that time was Hauptmann Walter Scheiner.

The emblem of 8./KG 2 that can be seen in the pennant in the previous photograph.

A Do 217 E-4 of 8./KG 2 showing the Staffel's distinctive emblem. (via Peter Taghon)

The aftermath of the accident of Do 217 E-4, Wk Nr 4372, which was coded U5+HT, at Deelen on 5 November 1942. The pilot on this occasion was Oberleutnant Josef Steudel, who, a highly decorated airman, would survive the war.

A low-level daylight flight as seen from the dorsal turret of a Do 217 E-4.

Oberleutnant Ernst Schneiderbauer's Do 217 E-4 of 3./KG 2, which had the Wk Nr 5441 and was coded U5+BL, having been camouflaged for participation in 'Pirate' attacks. One of Schneiderbauer's sorties was an attack on Poole which occurred on 16 December 1942. The following official account describes the events that day: 'At 1.23pm a Do 217 flying from northeast-southwest at 1,500ft machine gunned and dropped four 500kg bombs on the harbour area. A number of workshops and stores were demolished and partly demolished, and warehouses, workshops, stores, houses, shops and other buildings were damaged. One naval vessel was sunk and others damaged. Two persons killed, three seriously injured and 23 slightly injured. The Dorset Foundry factory, engaged on work of national importance, was damaged.'

A British intelligence report about the raid continued by saying: 'The Poole incident was directed against a small shipbuilding yard. Four 500kg bombs were dropped which the pilot thought straddled the target. This incident is confirmed from British sources from which it is learned that at 1.37pm three bombs were dropped on the quayside at Poole and one in the main harbour.'

Ernst Schneiderbauer pictured with his regular crew in March 1943. From left to right are Oberfeldwebel Gregor Eilbrecht (Bordmechaniker), Schneiderbauer himself, Oberfeldwebel Wilhelm Zacharias (Beobachter; injured on 8 February 1943), and Unteroffizier Hans Weber (Bordfunker). All but Zacharias would be taken prisoner on 12 March 1943.

Oberleutnant Ernst Schneiderbauer of 3./KG 2.

The last resting place of Schneiderbauer's Do 217 E-4, Wk Nr 5441, which crashed at Great Stainton on 12 March 1943. It was engaged in attacking Newcastle on the night of 11 March 1943 and fell victim to a Beaufighter of 219 Squadron flown by Flight Lieutenant John Willson.

The remains of one Do 217 that did not return. This is the still smouldering remains of Wk Nr 4382 which, flown by 4/KG 40's Unteroffizier Erich Dittrich, crashed in Bognor Regis on 16 December 1942. All on board were killed. It was his and his crew's first operational flight. The bombs from the Dornier were thrown from the bomber as it the ground, landing in Victoria Avenue and Hillsboro Road, mortally wounding 65-year-old Harold Booker and 26-year-old Georgina Hepton.

In the seconds before it crashed, at 14.47 hours Dittrich's Do 217 struck gas holder No. 2 at Bognor Gas Works, tearing open its side. At the time the raider was being pursued by Flying Officer Bill Cook and Flight Sergeant Len Warner in a 141 Squadron Beaufighter. Cook recalled the following: '[They] were flying at 180–200mph, I did not want to go below 150ft, to be at an advantage when the enemy climbed. Just before going over Bognor, the aircraft – still at 30–50ft – turned starboard. We followed, gaining, and he turned sharply to fly south, heading straight for the gasometer. He saw his starboard wing was likely to hit the gasometer and tried to correct his turn by throttling back and swinging to starboard. The wingtip, however, struck the gasometer, slewing him to strike the ground and catch fire as we flew over before we could fire.' (via Historic Military Press)

This is believed to be Do 217 Wk Nr 5444 of 6./KG 40, which force-landed at Soesterberg after combat on 18 December 1942. One source states that this was due to the engines being damaged in the engagement.

A Do 215 B-5 night fighter. In the background is Do 217 E-4 Wk Nr 1163, which later flew with 3./KG 2.

Do 217 E-4s of II./KG 2 training in 1942. The furthest aircraft is possibly Wk Nr 5465 U5+BM of 4./KG 2, which was shot down by either 219 Squadron or 416 Squadron on 7 July 1942. The aircraft crashed off the Dutch coast killing Feldwebel Johann Grandl and his crew. The nearer aircraft is definitely Wk Nr 5532 U5+FM of 5./KG 2. This was shot down by Wing Commander Rupert Clerke of 125 Squadron during an attack on Swansea. It crashed in the buildings at South Buckham Farm, Beaminster, Dorset, killing Staffelkapitän Hauptmann Hermann Euler and his crew. Amazingly the farmer, Arthur Swaffield, and his wife and son, who were all asleep at the time, were unharmed.

Looking at the nose and cockpit section of Do 217 J-1 Wk Nr 1251. It is possible this aircraft went to 4./NJG 3 as it is known that Wk Nr 1251 was damaged in an accident with this unit on 9 July 1942.

A side view of Do 217 J-1 Wk Nr 1251.

The funeral of Leutnant Gerold Buss, Unteroffizier Johann Brösamle and Prüfmeister (Maintenance Inspector) Fritz Tollkamp underway at Achmer aerodrome in Germany . Serving in 10./KG 2, they were killed when their Do 217 A-0, Wk Nr 2703 and coded U5+KU, crashed at Westerkappeln near Osnabrück on 30 March 1942.

Hauptmann Heinz Engel, Staffelkapitän of 5./KG 2 from October 1941 to January 1943, preparing for a flight. Engel would then command II./KG 2 until June 1944 and be awarded the Ehrenpokal and DKiG.

Hauptmann Heinz Engel resting below a Do 217 E-4 coded U5+SN, possibly Wk Nr 4247. The latter crashed on landing at Eindhoven on 15 June 1942. It had been returning from a mine-laying sortie off the Isle of Wight. Oberfeldwebel Fritz Hilgenfeldt and his crew were all killed.

Do 217 bomber aces from II./KG 2 seen at Eindhoven in May 1942. Left to right they are: Hauptmann Alfred Kindler (Staffelkapitän 6./KG 2, who became a prisoner of war on 31 July 1942), Oberleutnant Friedrich Dörflinger (6./KG 2, killed on 30 July 1942), Oberleutnant Wolfgang Hankammer (5./KG 2, killed on 14 January 1945), Oberleutnant Karl-Ludwig Krüger (6./KG 2, killed 19 August 1942), Leutnant Ernst Andrew (5./KG 2, killed 12 February 1945) and Leutnant Anders Lembcke.

The wreckage of Dornier Do 217 E-4 Wk Nr 1225, coded U5+AD, that that was shot down by Flight Lieutenant Lindsay Black, at the controls of a Spitfire of 485 Squadron, on 3 August 1942. At the time the bomber was part of an attack on Wellingborough. It crashed near Cranford, a small village on the outskirts of Kettering, killing all four crew members, Unteroffizier Eugen Beyerer, Unteroffizier Erich Huwald, Unteroffizier Zerak and Gefreiter Frisch.

Do 217 E-4 U5+HP of 6./KG 2. The only aircraft recorded as having this code was Wk Nr 4221, which crash-landed near Abbeville, due to fighter damage, on 19 August 1942.

A Do 217 E-1 of II./KG 2 pictured in 1942.

The pilot's position in a Do 217 E-4.

A view of the Beobachter's position in a Do 217 E-4.

Looking down into the Beobachter's position in a Do 217 E-4.

The Bordfunker's position directly below the turret in a Do 217 E-4.

Repair and maintenance work underway in the cockpit of a Dornier Do 217, Germany, 1942.

Another view of repair and maintenance work underway in the cockpit of a Dornier Do 217, Germany, 1942.

An engine exchange underway at a Luftwaffe base in the Netherlands, May 1942.

Groundcrew at work on Do 217s in the Netherlands, May 1942.

A hoist is used to manoeuvre a Do 217 engine in the Netherlands, May 1942.

The replacement engine is offered up to the waiting Do 217.

Another shot of an engine exchange on a Do 217 underway in the Netherlands, May 1942.

Another shot of an engine exchange on a Do 217 underway in The Netherlands, May 1942.

Mechanics at work installing a propeller on a Do 217 at Schiphol in May 1942.

Groundcrew working on a Do 217.

Luftwaffe mechanics work on propeller of a Dornier Do 217 at Schiphol, May 1942.

Repair and maintenance on a Dornier Do 217, Germany, 1942.

This Do 217 N-2, which had the Wk Nr 0174 and was coded PE+AW, had been converted from a Do 217 E-1. Note that there is no radar is fitted.

This Do 217 N-0 is believed to be the sixth aircraft. It is fitted with FuG 202 radar.

Looking down on a Do 217 N-0 with the codes GG+YG – which, in this instance, was the seventh example of the variant.

The seventh Do 217 N-0, GG+YG, as seen from the opposite angle.

Reputed to be Hauptmann Rudolf Schoenert's personal Do 217 J, this aircraft clearly shows the radar and gun arrangement in the nose. Schoenert would be awarded the Ritterkreuz mit Eichenlaub and survived the war. Between 16 August and 15 October 1943, he claimed at least twenty-five aircraft over the Eastern Front flying Do 217s with I./NJG 100. His final score as a night fighter pilot was sixty-five.

The first of three views of the Do 217 K-01 Wk Nr 4401, coded KE+JA.

Looked straight at the nose of Do 217 K-01 Wk Nr 4401. This aircraft was one of an initial batch of ten pre-production aircraft.

Leutnant Hartmut Holzapfel's first crew on 6./KG 2. They are Unteroffizier Friedrich Vahland (Bordfunker; wounded 26 October 1941), Feldwebel Gerhard Dörr (Bordmechaniker; became a prisoner of war on 15 March 1943), and Oberfeldwebel Josef Schuster (Beobachter; killed on 5 February 1942).

Damage caused on the ground to a Do 217 E of KG 2. The date and location is not known.

Loading a 250kg bomb to a Do 217 E-1 of II./KG 2.

Chapter 3
The Dornier Do 217: 1943

With the Wk Nr 4381 and coded F8+BM, this Do 217 E-4 was operated by 4./KG 40. It hit the ground at Fairlight, East Sussex, on 4 January 1943. Feldwebel Hartmut Eucker and his crew were all killed.

Flying with Flugbereitschaft des Oberbefehlshaber Süd, this Do 217 K-1, Wk Nr 4463, was reported to have suffered 50 per cent damage at Trapani, Italy, on 11 February 1943. It would appear that it was repaired as it is shown here while with Flugzeugführerschule (C) 5 at Anklam in Germany.

Another view of Do 217 E-4 Wk Nr 4272 of 9./KG 2. It was lost on 15 January 1943, by which time it had been given new codes of U5+AT. Unteroffizier Hans Unglaube and his crew were killed, as was the Staffelkapitän, Oberleutnant Franz Holtz. It is possible that U5+AT was shot down over the North Sea by Flight Lieutenant Joe Singleton of 25 Squadron.

A Do 217 K of KG 2 taxies out on to a wintery runway in January 1943.

Do 217 K-1 Wk Nr 4406 was flying with KGzbV 21 when it suffered a minor accident at Schwäbisch Hall on 21 January 1943.

An unidentified Do 217 K-1 with Maander camouflage undergoing maintenance in a hangar. The Wk Nr on the nose cannot be discerned.

Damage to Leutnant Hermann Walther's Do 217 E-4 in February 1943. Walther, seen here on the right, flew with III./KG 2.

Another view of the damage sustained by Leutnant Hermann Walther's Do 217 E-4 in February 1943.

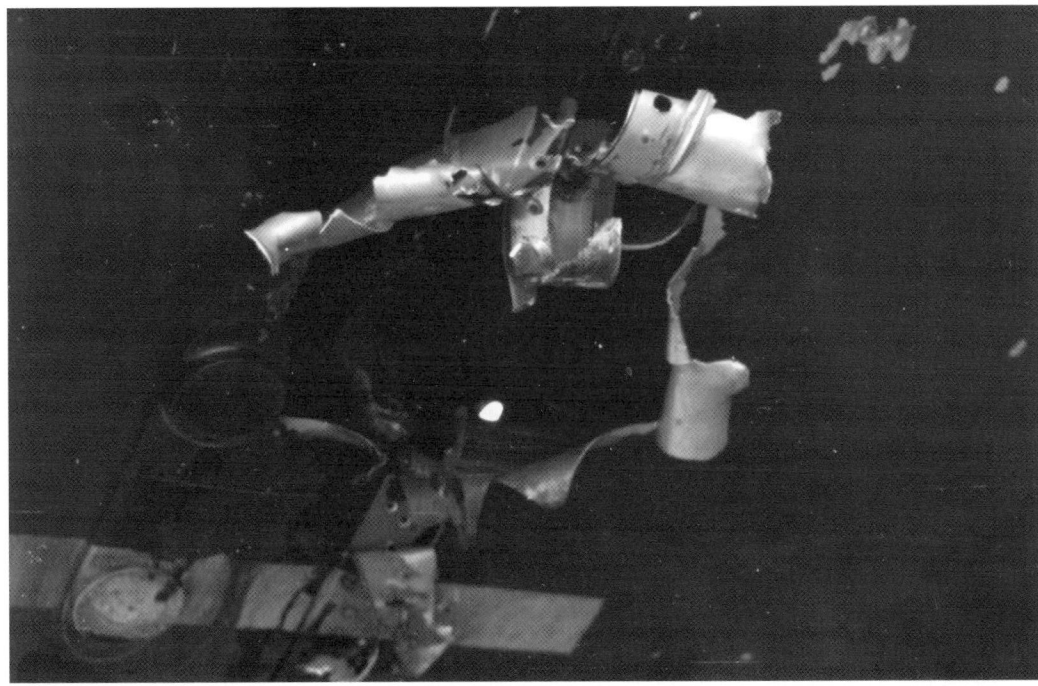

Do 217 E-4 U5+IR of 7./KG 2. Despite the last two numerals of the Wk Nr being visible on the nose, the fate of this aircraft has not been ascertained.

An air-to-air shot of Do 217 E-4 U5+KN of 5./KG 2. While returning from an attack on Swansea on the night of 16 February 1943, Wk Nr 1175 U5+KN suffered engine failure and was destroyed in a crash near at Sainte Croix Hague near Cherbourg. Feldwebel Walter Schwarz and two crew were killed. A later U5+DN, Wk Nr 4239, was lost on 17 August 1943.

Almost certainly from the same series as the previous image, this photograph shows Do 217 E-4 U5+ON, the letters ON appearing on the tail. The only record of an aircraft with this code is Wk Nr 4228, which was lost on the night on 29 May 1942.

The fuselage band on this Do 217 E-4 indicates that it belonged to a Nachtaufklärungsstaffel operating on the Eastern Front.

Two views of a Do 217 E-4 that is believed to be from 4.Nachtaufklärungsstaffel. (via Del Davis)

The aftermath of a crash-landing by an unidentified Do 217 K-1 of 4.Nachtaufklärungsstaffel.

In the early hours of 8 March 1943, Squadron Leader Geoff Goodman and Flying Officer Bill Thomas, at the controls of a 29 Squadron Beaufighter, intercepted a Dornier Do 217 in the skies over West Sussex. This was Do 217 E-4 Wk Nr 5526, which was coded U5+EH. Flown by Feldwebel Gunter Vestewig, it fell to earth at Vann Common, near the village of Fernhurst – where the soldier seen here is examining some of the wreckage. The subsequent RAF intelligence report notes that the 'aircraft appears to have become thoroughly lost over the Channel on its way to the target and when the crew crossed the Sussex coast they thought they were well west of Selsey Bill. The pilot was flying at about 11,000ft making no attempt to evade night fighters by change of height or course. A warning was suddenly received that a night fighter was in the neighbourhood but before evasive action could be taken an attack developed. Hits were scored in the starboard fuel tanks and in the starboard engine and when the latter burst into flames orders were given to bale out.' (Andy Saunders)

Another view of the wreckage of Vestewig's Do 217 E-4 at Fernhurst. Another report states that local witnesses had observed that 'the machine was on fire in the air. It dived into the ground vertically and most of the wreckage was buried and the rest badly burnt.' It went on to add that 'bullet strikes were traced in the wreckage', on the leading edge of a propeller blade, through the fuselage forward of the tail wheel, and in various places in the tail unit. (Andy Saunders)

A final photograph of the wreckage of Do 217 E-4 Wk Nr 5526. Vestewig and his Beobachter, Obergefreiter Gerhard Polzin, both baled out successfully and quickly became prisoners of war. The Bordfunker, Obergefreiter Hans Witkopp, followed them out, but without his parachute and fell to his death. Of the fourth crew member, Obergefreiter Franz Huske (the Bordschütze) the RAF report notes that 'the aircraft crashed in flames and burned fiercely on the ground and it seems probable that his body is in the wreckage'. It was only some years after the war that his remains were found at the crash site. (Andy Saunders)

The bola hatch can be seen in position on this Do 217 E-4, that with the Wk Nr 5332. This aircraft had a long career, starting as a E-2 with II./KG 2. It was damaged in an accident on 19 December 1942. Repaired and upgraded to a E-4, it was damaged in combat on 13 March 1943, before suffering another accident on 3 April 1943. Once more repaired, it was damaged in another accident with IV./KG 2 on 4 October 1943. Taken at Schiphol, this photograph is dated February 1943. (via Peter Taghon)

Although seen here while being operated by KG 2, this Do 217 K-1 is believed to be that with the Wk Nr 4484. Being operated by Flugbereitschaft Luftflottenkommando 2, with the code letters F5+IH, it was lost on 19 March 1943, when it crashed near the airfield at Tunis. Unteroffizier Willi Lock, two crew and three passengers were killed.

This photograph was taken at Coulommiers in the spring of 1943 when II./KG 2 was in the process of converting to the Do 217 M-1.

One of II./KG 2's Do 217 M-1s, again pictured at Coulommiers in the spring of 1943. U5+HS is possibly Wk Nr 722851, which was lost attacking Plymouth on the night of 15 November 1943.

The nose section of U5+HS. At the time of its loss, this aircraft was being flown by Leutnant Klaus Dicke and his crew. No RAF claim was made, and it is presumed that the aircraft came down in the Channel. Dicke and his crew remain listed as missing.

A Do 217 K-1 that is presumably from KG 2, though there are no markings that would help confirm its unit.

A Do 217 K-1 of Flugbereitschaft Luftflottenkommando 2, this being indicated by the badge on the nose and the white Mediterranean fuselage band. Do 217 K-1, with the Wk Nr 4601 and coded F5+NH, was reported missing between Rome and Tunis on 22 April 1943. It was a possible victim of friendly fire.

A Do 217 E-4, that coded H, seen at Stavanger-Sola in April 1943. It is almost certain that it was at Sola to participate in the devastating attack on Aberdeen on the 21st of that month. In this, a force of some twenty-five Do 217s, most of which were normally based in the Netherlands, had been forward-based to Sola. It was from there that the raiders struck out across the North Sea. In the raid, some ninety-eight people were killed, as well as twenty-seven soldiers in their barracks. A further 93 civilians were seriously injured and 141 slightly wounded. More than 9,300 buildings were damaged.

Do 217 K-1 Wk Nr 4437. This aircraft was reported as suffering minor damage at Salon on 22 April 1943, while with IV./KG 2. Note the letter E on the leading edge.

Major Walter Bradel, second from left, in front of Do 217 K-1, coded U5+AA, of Stab/KG 2. Some records state this aircraft was Wk Nr 4415, the aircraft in which Bradel was flying when he was killed on 5 May 1943. Other sources, however, claim that it was Wk Nr 4405, which suffered engine failure while with 2.Nachtaufklärungsstaffel on 22 October 1943, causing it to crash in Belarus. On the far left is Major Gottfried Buchholz, Geschwader TO.

A pair of Do 217 E-4s of 9./KG 2. By the middle of 1943, III./KG 2 was converting to the K and M variants.

Photographed in May 1943, this Do 217 E-4, coded F8+CM, was being operated by 4./KG 40. On 18 May 1943, the Do 217, coded F8+CM, and which had the Wk Nr 5424, failed to return from operations. Oberfeldwebel Erich Harms and his crew were reported missing, believed killed.

Leutnant Wolfgang Nestler, of 6./KG 2, poses for the camera in front of Do 217 K-1, coded U5+BP. Though he had an eventful war, eventually flying the Ju 188 with KG 2, Nestler would survive to see the end of the fighting in Europe.

Dornier 217 E-4 U5+DV of 11./KG 2 pictured being recovered after suffering a landing accident in 1943. (via Olrogg)

Another KG 2 Do 217 E-4 has come to grief and suffered catastrophic damage to its fuselage and tail. The date and location of this incident is not known. (via Olrogg)

A Do 217 E-5 with the bulge for the Kehl guided bomb directional equipment clearly visible on the side of the cockpit.

A close-up of the nose of a Do 217 N-1; the last three digits of the Wk Nr 174, though visible on the nose do not give any clue to the aircraft's fate. The FuG 202 Lichtenstein aerials, four 20mm cannon and two MG 81s can clearly be seen.

This Do 217 J-1, coded G3+PV, of II./NJG 101, was photographed at some point in 1943. Note that its aerials have been removed. (via Peter Taghon)

Another of II./NJG 101's Do 217 N-1s, this time coded G3+SV, also pictured in 1943. (via Peter Taghon)

A member of air or ground crew is pictured leaning on the tail of a Dornier 217 N-2 of NJG 101, which had the code letters G3+NO, at Sofia-Bojuritsche airfield in 1943. The *Schräge Musik* cannon can be seen on the top of the fuselage behind the cockpit.

Another picture of G3+NO at Sofia-Bojuritsche airfield in 1943. This time we are presented with a clear view of the tail braking parachute.

A dramatic air-to-air shot of the cockpit area of a Do 217 K-1 of KG 2. It is believed to be Wk Nr 4425, which was flying with II./KG 2 in August 1943.

Battle damage to the tail of Leutnant Ernst-Karl Fara's Do 217 of 1./KG 2, 1943. Fara flew his first mission with KG 2 in June 1943, but by the end of the year was flying Ju 188s with I./KG 66.

Combat damage to a the tail of a Do 217 E-4 of 8./KG 2. It has not been possible to date this and the following photographs.

A second picture of combat damage to a Do 217 E-4 of 8./KG 2.

Do 217 E-4, Wk Nr 4285, coded U5+EW, of 12./KG 2, pictured having suffered an accident at Villaroche on 17 July 1943.

A member of groundcrew cycles past a Do 217 E-4 of I./KG 66 at Chartres in the summer of 1943. Note the additional Y Gerät aerial behind the cockpit. (via Heikop)

Another of I./KG 66's Do 217 E-4s at Chartes in the summer of 1943. (via Heikop)

Do 217 K-1 Wk Nr 4464, coded 3E+HZ, after an accident at Montdidier, France, at 00.30 hours on 26 June 1943. (via Heikop)

A closer view of the nose of Do 217 K-1 Wk Nr 4464, after the accident at Montdidier on 26 June 1943.

Do 217 K-1 Wk Nr 4472 of I./KG 66 running up its engines at Chartres, 1943.

Personnel of I./KG 66 pictured at Chartres while undergoing training in the spring of 1943. (via Heikop)

A photograph of the starboard side of the Do 217 E-4, coded Z6+OH, of 1./KG 66, at Montdidier during 1943. (via Heikop)

A member of the crew of Do 217 E-4 Z6+OH, can be seen entering or exiting their aircraft at Montdidier during 1943. (via Heikop)

A close-up of Do 217 E-4 Z6+OH's tail section, this being a third photograph taken at Montdidier in 1943. (via Heikop)

Another Do 217 that was pictured at Montdidier in 1943 – in this case an M-1 variant which, with the Wk Nr 126 and codes Z6+AK, was being flown at the time by 2./KG 66. (via Heikop)

Three Do 217 M-1s of 1./KG 2, those coded U5+AH, U5+CH and U5+DH, photographed from another of the unit's aircraft while airborne during the summer of 1943. With the Wk Nr 40705, U5+AH would be lost during an attack on Plymouth on 12 June 1943 while being flown by Staffelkapitän Hauptmann Otto-Wilhelm Pöhler. His Do 217 was the only loss.

Taken during the same flight as the previous image, this is a closer view of 1./KG 2's U5+CH, which had the Wk Nr 40728. It was lost during an attack on London on the night of 3 February 1944. It was being operated by 8./KG 2 at the time. (via Peter Taghon)

Originally thought to have been a Do 217 E-4 of I./KG 66, the lack of Y Gerät aerials would indicate otherwise. It is believed that this Mäander pattern camouflage was adopted by II. and III./KG 2 sometime during 1943.

Crewmen of a Do 217 K-2 of 8./KG 100 pictured at Istres during August or September 1943.

Also seen at Istres during August or September 1943, these Do 217 K-2s are from 7./KG 100.

The Dornier Do 217: 1943

The first of a series of propaganda photographs – five of which we include here – relating to Major Bernhard Jope and various members of groundcrew from III./KG 100, 1944.

Major Bernhard Jope in discussion with groundcrew working on the wheels and undercarriage of one of his unit's Do 217s. Jope joined the Luftwaffe in 1935, eventually being posted to II./KG 253 at Nordhausen where he was the Gruppe TO. Promoted to Oberleutnant in June 1938, the following January saw Jope flying in the latter stages of the Spanish Civil War, during which he flew about twenty missions and was awarded the Spanish Cross in Bronze with Swords.

In May 1939, Jope was posted to KG 76 as the TO and an instructor, and flew operationally in Poland after which he was posted to KG 28 this time as the Geschwader TO. However, his time with KG 28 was short as in the summer of 1940, he was posted to the newly formed I./KG 40, joining as a pilot with 2.Staffel and acting as the Gruppe TO.

Jope was awarded the EK I in September 1940. At the end of the following month he carried out a mission which saw him immediately being regarded as a first class Focke-Wulf Fw 200 Condor pilot. Having taken off from Bordeaux on 26 October 1940 for an armed reconnaissance over the Atlantic, he located a 42,500 GRT liner 140km west of the Isle of Arran. This was the liner *Empress of Britain*, which was being used as a troop transporter – at the  time she was Britain's second largest ship and the tenth largest merchant vessel in the world.

Without hesitation, Jope attacked. He dropped six 250kg bombs, two of which hit the target. The liner was crippled and caught fire; the German crew heard the S.O.S. message being transmitted.

Jope's Fw 200 was lightly damaged by anti-aircraft fire, so having reported what he had achieved, he turned for home. The liner was mortally hit, with at least twenty-five merchant seamen killed. She was taken into tow but to no avail; the liner's location had been transmitted to *U-32*, which sank the ship two days. Jope's attack was a propaganda coup for the Luftwaffe and was widely reported. For his achievements, Jope was awarded the RK.

In April 1941, Jope was given command of 3./KG 40 with the rank of Hauptmann. He continued to fly missions well into the summer of 1942, but after the second of his two brothers was killed in action he was posted to Rechlin for service trials on the He 177.

His rest from operations was not long as, in May 1943, he took command of IV./KG 100 in preparation for operations in which Dornier Do 217s would carry remotely controlled glide bombs for anti-shipping missions. At the end of July 1943, Jope took command of III./KG 100 and, operating from southern France, carried out glide bomb missions over the Mediterranean with great success.

In September 1943, Jope was given command of KG 100, a position he held until August 1944 during which time he was awarded the EL and promoted to Major. From then on, he commanded KG 30 for a short time and at the end of the war was serving as a staff officer.

It was Jope who led the attack in which crews from III./KG 100, operating from Marseilles-Istres, sank the Italian battleship *Roma* and severely damaged *Italia* west of Corsica on 9 September 1943.

Jope talking to some of his groundcrew. He was awarded the Eichenlaub in March 1944 by which time he had moved from being Gr Kdr III./KG 100 to Kommodore KG 100.

The Dornier Do 217: 1943 159

A close-up of Major Bernhard Jope. He is seen wearing the Ritterkreuz mit Eichenlaub.

This Do 217 K-1 with the Wk Nr 4518 is reputed to have been photographed at Bordeaux and was an aircraft intended for Japan. Wk Nr 4519 was reported to be on the strength of II./KG 2 in July 1943.

Though quite dark, this photograph shows the aftermath of Captain Aramis Ammanato's crash landing of a Do 217 J of 235a Squadriglia in 1943. The Regia Aeronautica ordered six Do 217 J-1 and six J-2s (Giuseppe Grande, via Bussi)

Captain Ammanato's Do 217 as seen from the front. Ammanato commanded the 235 Squadriglia and made the only claim for Italian Do 217 night fighters, a Lancaster of 207 Sqn, near Milan, 16 July 1943 (Giuseppe Grande, via Bussi)

A Do 217 M-1 of KG 2 with the crew of who is believed to be Leutnant Arendt, the identity of whom and fate has not been ascertained. (via Peter Taghon)

Do 217 M-1 U5+ET of 9./KG 2 photographed at some point during the summer of 1943. Two Do 217s with this code were lost that year.

On the night of 15/16 August 1943, one of the two Do 217 M-1s, coded U5+ET, that with the Wk Nr 722852, crashed north of Portsmouth, having been despatched to attack the city. Caught in a searchlight beam, the pilot, Leutnant Theo Bach, took violent evasive action which resulted in the aircraft slamming into the ground at Broadway Farm, near Lovedean in Hampshire at 00.10 hours.

Another view of the badly disintegrated wreckage of Do 217 M-1, Wk Nr 722852, at Broadway Farm. All four members of Bach's crew were killed. Only the flight engineer, Obergefreiter Werner Neubert was identified and buried. The second U5+ET had the Wk Nr 56162; it was shot down off Ramsgate on the night of 15 September 1943 by Flying Officer Ron Watts of 488 Squadron.

Brand new Do 217 Es pictured while ready for delivery. Second from the left is Do 217 E-2, Wk Nr 1145, RH+ES. Having been upgraded to E-4 standard, it suffered 50 per cent damage when a tyre burst whilst landing at Vannes on 11 August 1943. By then it was U5+ZN of 5./KG 2.

Do 217 E-2 RH+ES can also be seen on the left in this photograph. To the right, meanwhile, is the E-2 with the Wk Nr 1115 and coded RE+CP. It suffered 30 per cent damage at Rennes, while flying with II./KG 2, on 2 July 1943. Then, as U5+DP of 6./KG 2, it crashed 5km north-west of Utrecht. Unteroffizier Reinhard Elster and Unteroffizier Will Laux were both injured.

A Hs 293 mounted under the wing of a Do 217 E-5, the latter from either Lehr und Versuchsstaffel 293 or Erprobungskommando 15.

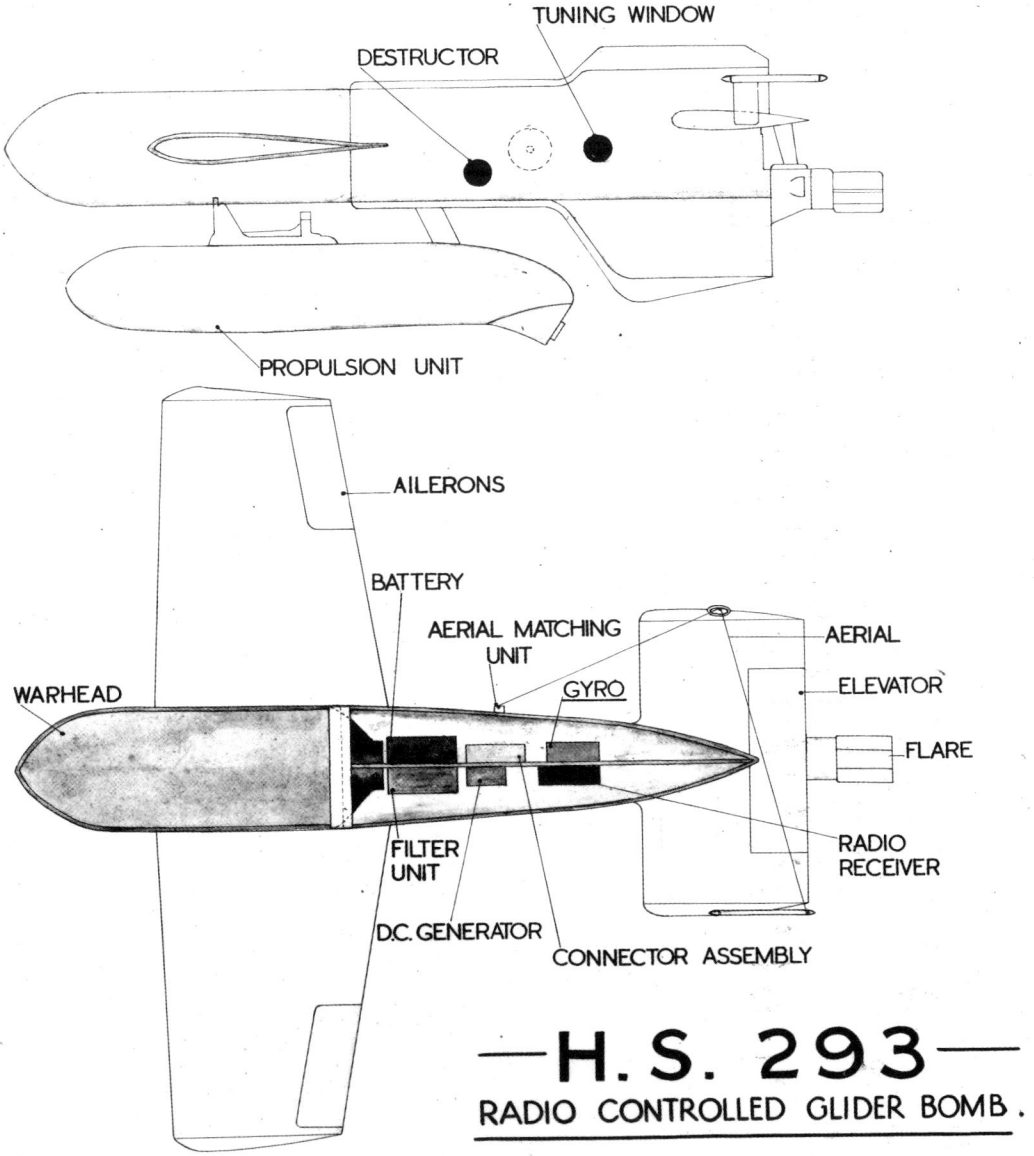

Produced by the RAE in February 1946, this schematic shows the simplicity and effectiveness of the Hs 293.

An example of a PC 1400X, which was more commonly known as Fritz X.

The Beobachter steers a Hs 293 or Fritz X to the target. The Hs 293 was intended to destroy unarmoured ships, unlike the unpowered, armour-piercing Fritz X.

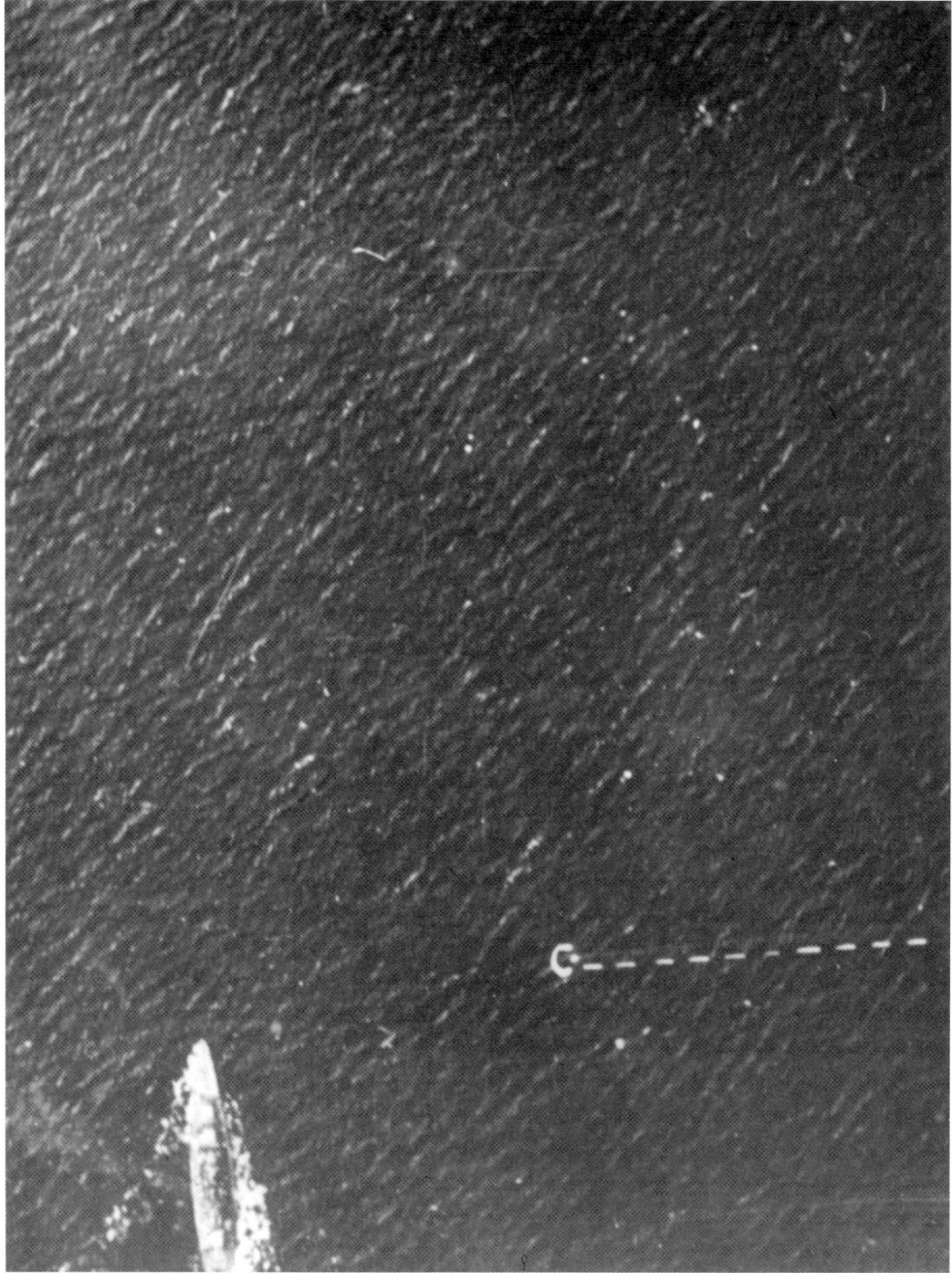

On 9 September 1943, a day after the proclamation of the 1943 Italian armistice, the Italian battleship *Roma*, along with many of the principal units of the Italian fleet, sailed from La Spezia. Admiral Bergamini had been ordered to leave La Spezia in order to prevent the fleet from falling into German hands and reach Allied-controlled ports. The various warships were, however, attacked by Dornier Do 217s from III./KG 100, the latter operating from operating from Marseilles-Istres. *Roma* was soon targeted by the Luftwaffe crews – as seen here.

Having shadowed the Italian fleet for some time, during the afternoon of 9 September 1943 the Dornier Do 217s made their move. At least two Fritz X guided missiles hit *Roma*, the second at 16.02 hours, causing fatal damage.

The final act in the sinking of *Roma* came when the No. 2 15-inch turret was blown over the side by a massive explosion, this time from the detonation of that turret's magazines – it maybe this event that was captured in this photograph. This resulted in catastrophic flooding, and the battleship began to go down by the bows. It eventually capsized and broke in two.

Another victory for the Dornier Do 217 crews – this is the moment that a Fritz X struck the light cruiser USS *Savannah* (CL-42) on 11 September 1943. At the time, *Savannah* was supporting Allied forces ashore during the Salerno landings. The guided bomb hit the top of the ship's No. 3 6/47 gun turret and penetrated deep into her hull before exploding. The photograph shows the explosion venting through the top of the turret and also through *Savannah*'s hull below the waterline. (USNHHC)

Pictured while USS *Savannah* was undergoing emergency repairs off Salerno, sailors spray water onto the hole caused by the Fritz X that hit the light cruiser's No. 3 6/47 gun turret on 11 September 1943. Note the life rafts atop the turret, one of which has been cut in two by the bomb. Also note the turret's armoured faceplate. This view was taken looking forward, with the No. 2 6/47 gun turret in the immediate background. The original caption, released on 2 November 1943, reads in part: 'A round, clean hole marks the point of entry of a Nazi bomb on the cruiser *Savannah*. Inside, all was chaos, smoke, blood, and death.' (National Museum of the US Navy)

During operations around the Salerno landings, the light cruiser USS *Philadelphia*, seen here early in 1943, narrowly avoided being hit by two Fritz Xs on 13 September that year; one came within 100 yards, the other within 100ft. At 14.40 hours the same day the light cruiser HMS *Uganda* was conducting close fire support missions when she suffered a direct hit from a Fritz X dropped from a Do 217 that was never seen. The bomb penetrated through seven decks, before bursting through the bottom of the hull to explode under the keel. Sixteen men lost their lives. A US Navy tug was able to tow *Uganda* to safety, but she was out of action for many months. Other British destroyers suffered near-misses from guided bombs. (USNHHC)

A German photograph taken during the attack on the Royal Navy battleship HMS *Warspite* by three Do 217s of KG 100 on 16 September 1943. Of the three Fritz Xs launched at the battleship, one struck near the funnel, cutting through her decks and making a 20ft hole in the bottom of her hull, while a second near-miss ripped open the torpedo bulges. The third device missed the warship. Although the damage was considerable, HMS *Warspite*'s casualties amounted to only nine killed and fourteen wounded.

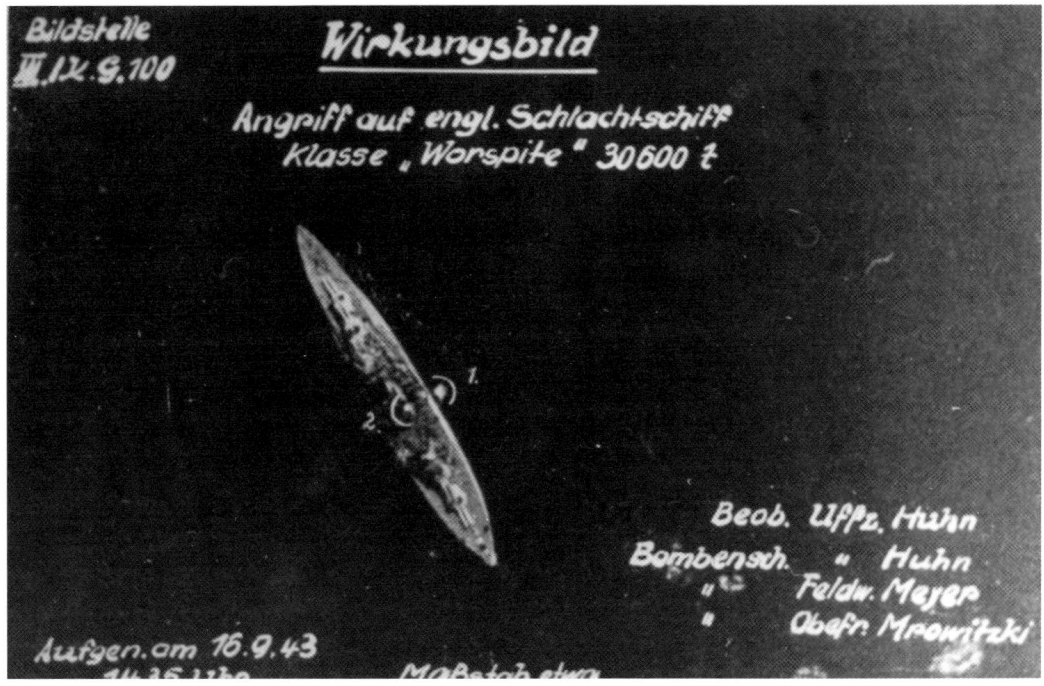

The Dornier Do 217: 1943   175

A dramatic photograph showing the moment that a crew member baled out of the Do 217 that was being shot down by Flight Lieutenant Roy Hussey of 72 Squadron, in the Ascerno/Cipriano region, on 19 September 1943. This is possibly the Do 217 K-2 of 7./KG 100 flown by Leutnant Karl-Ernst Michelis.

The remains of this Do 217 K-2 of III./KG 100, which had the Wk Nr 4560, were discovered at Foggia in Italy after the Germans had retreated.

This picture of the rear section of Do 217 K-2 Wk Nr 4560 provides an excellent view of the Giesskanne in the aircraft's tail.

Dornier Do 217s of IV./KG 2 pictured during a training flight. Furthest from the camera is the E-4 which, coded U5+OU, was flown by Hauptmann Fritz Hauptmann. The nearest aircraft is U5+AU, on this occasion flown by Unteroffizier Hermann Heller. The photographer was Feldwebel Kurt Pöggeler in U5+JU. Hauptmann took command of 7./KG 2 on 21 February 1944, only to be killed in action attacking Portsmouth on the night of 26 April 1944. Heller would fly with 2./KG 2; both he and Pöggeler survived the war.

Photographed in October 1943, this Do 217 E-4, coded E, is reputed to be from I./KG 66, but, once again, has no additional aerial. What can be seen are FuG 216 Neptun R radar aerials under the wings.

Groundcrew and armourers bomb up the Do 217 E-4, coded E, that is believed to be from I./KG 66.

Do 217 E-5 Wk Nr 5663 KE+GG later flew with II./KG 100. Note the bulges on the nose for the guidance equipment.

Do 217 E-5 Wk Nr 5554, again of either Lehr und Versuchsstaffel 293 or Erprobungskommando 15. While serving in I./KG 100, this aircraft was reported as having been damaged on 13 September 1943.

This Do 217, believed to be an E-5 from II./KG 100, was photographed from a Ju 88 C-6 of V./KG 40.

The number on the nose of this Do 217 K-2 indicates that it had the Wk Nr 4572. This aircraft was reported damaged with 5./KG 100 at Kalamaki, on the south coast of Zakynthos (Zante), a Greek island, on 15 November 1943. On that day, a force of forty-five USAAF North American B-25 Mitchells, escorted by thirty-six Lockheed P-38 Lightning bombed the Luftwaffe airfield there, damaging runways, dispersal areas and hangars, as well as some of the aircraft based there. Large fires and explosions were reported by the attackers.

Another Do 217 of 5./KG 100 at Kalamaki on 15 November 1943, was that seen here – E-4 Wk Nr 1148. In this case, the aircraft was totally destroyed.

The nose of a Dornier Do 217 on a snow covered airfield in the Netherlands, 1943.

A side view of Do 217 K-01 Wk Nr 4401.

This Do 217 M-1 is believed to be that with the Wk Nr 56013 and which carried the codes DN+UE.

A side view of the Do 217 M-1 which is believed to be that with the Wk Nr 56013.

The cockpit of a Do 217 M-1.

An external view of the cockpit of a Do 217 M-1. Note the central, reinforced strip that was added as protection against barrage balloon cables.

An unidentified Do 217 E-4. The pennant would indicate a Staffelkapitän or other senior officer, but, aside from an Iron Cross, the writing below cannot be determined.

A Do 217 E-4, presumably from II./KG 40, showing hastily applied night camouflage.

What appears to be a Do 217 M-1 testing the Rüstsatz N.25 Bänderfallschirm-Sturzflugbremse (ribbon parachute for dive braking).

Another view of what may be a Do 217 M-1 testing the Rüstsatz N.25 Bänderfallschirm-Sturzflugbremse (ribbon parachute for dive braking).

Chapter 4
The Dornier Do 217: 1944–1945

A camera gun still of the attack made by Pilot Officer John McLaughlin of 609 (West Riding) Squadron at Gilze-Rijen on 4 January 1944.

An unidentified Do 217 K-1 of 4.Nachtaufklärungsstaffel pictured after a crash-landing on the Eastern Front in 1944. (via Del Davis)

This Do 217 M-1, Wk Nr 40703 and coded U5+CL, suffered minor damage returning from a mine-laying sortie on 4 September 1943. Feldwebel Willi Sprinzing and his crew were all uninjured. This aircraft was lost over France in January 1944 while returning from an attack on London. (via Andy Saunders)

Two 4./KG 100 Do 217 crewmen, Unteroffizier Paul Hoffmann and Leutnant Dr Franz Lutz, pictured together while at the controls of a Heinkel He 111. (via Urbanke)

Do 217 N-1, coded NH+CM. It is possible to make out the numbers 91 on the nose. A Do 217 N-1 with the Wk Nr 51491 was serving with III./NJG 1 on 1 February 1944 when it was damaged in an accident. The same aircraft was being operated by 7./NJG 4, with the code 3C+HR, when it was destroyed in a crash near Consenvoye, 10 miles north-west of Verdun, on 2 April 1944. Oberfähnrich Siegfried Voigt and his two crew were all killed. (via Peter Taghon)

Engine runs being carried out on a Do 217 N-1 of an unidentified unit. (via Peter Taghon)

A Do 217 N-1, which was coded W7+RP, of 6./NJG 101 photographed in 1944. Note the different coloured cowlings. (via Peter Taghon)

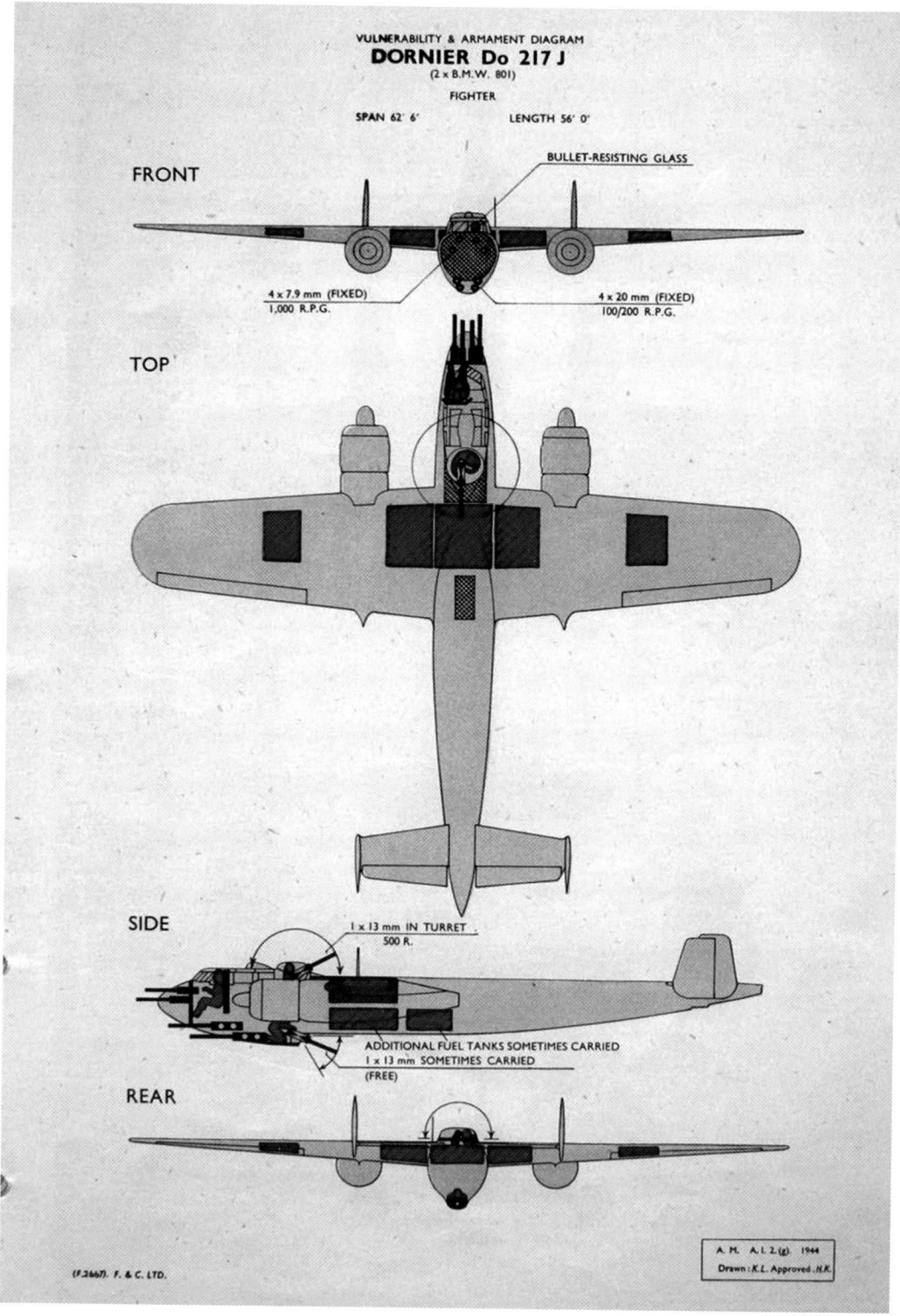

This RAF 'vulnerability and armament' recognition guide to the Do 217 J was produced in 1944. (via Dave Wadman)

These photographs of a Do 217 E-5, coded +HN, of 5.Staffel have always been a puzzle. Initially thought to be I./KG 66, this Geschwader did not have a II Gruppe and the aircraft lacks a Y Gerät aerial. Note that this aircraft does not have the E-5's Kehl bulge in the nose.

Another view of the Do 217 E-5, coded +HN, of 5 Staffel. This camouflage was used by II./KG 100, and a Do 217 E-5 Wk Nr 5631, coded 6N+HN, was lost in the Anzio area on 21 January 1944.

The nose of a Do 217 N-1, the unit of which has not been established, with the last two digits of the Wk Nr, radar antenna and weaponry all visible. (via Peter Taghon)

This Do 217 N-2 Wk Nr 1570 and coded 3C+IP, of 6./NJG 6 became lost on a training sortie from Taveaux on 2 May 1944. It subsequently landed at Basel-Birsfelden in Switzerland.

The tail of Do 217 N-2 Wk Nr 1570, which, as can be seen here, was coded 3C+IP. Having landed at Basel-Birsfelden, the aircraft was interned by the Swiss authorities, as was the crew of Feldwebel Günther Konzac (pilot), Unteroffizier Arthur Ruprecht (Bordfunker) and Obergefreiter Alfred Elster (Bordschütze). The aircraft was not returned and was scrapped after the war.

Do 217 M-1s of 4.Nachtaufklärungsstaffel during 1944. The aircraft furthest from the camera is that coded K7+XM. (via Del Davis)

Taken during 1944, this photograph of a Do 217 M-1 of 4.Nachtaufklärungsstaffel shows the twin MG 81Z machine guns in the nose and the cartridge ejection chute snaking beneath. (via Del Davis)

Luftwaffe groundcrew in front of a Do 217 M-1 of 4.Nachtaufklärungsstaffel, 1944. (via Del Davis)

Groundcrew working on Do 217 M-1 K7+CM of 4.Nachtaufklärungsstaffel. With the Wk Nr 426343, Do 217 M-K7+CM was reported missing off Finland on 12 June 1944. Unteroffizier Johannes Glausch and his crew were all reported missing. (via Del Davis)

Formation flying practice for Do 217 M-1s of 4.Nachtaufklärungsstaffel, 1944. (via Del Davis)

The Do 217 J-2 that was used by General Josef Kammhuber of Luftflotte 5. This aircraft was coded N9+AA of Kurierstaffel Norwegen, and also reported as carrying codes CG+KW. It is believed this photograph was taken in March 1944 at Alakurtti, Finland, during an inspection of Nachtschachtgruppe 8.

The Do 217 M-1 Wk Nr 56125 that was coded U5+UK. This aircraft was damaged in an accident at Eindhoven on 7 September 1943. Repaired, it became U5+CT of 9./KG 2. It was reported missing during an attack on Hull on the night of 20 April 1944; Obergefreiter Hermann Wendt and his crew were all killed.

This shot of Do 217 K-2 Wk Nr 4572 shows its increased wingspan. This aircraft was reported as being damaged while with 5./KG 100 at Kalamaki, Finland, on 15 November 1943.

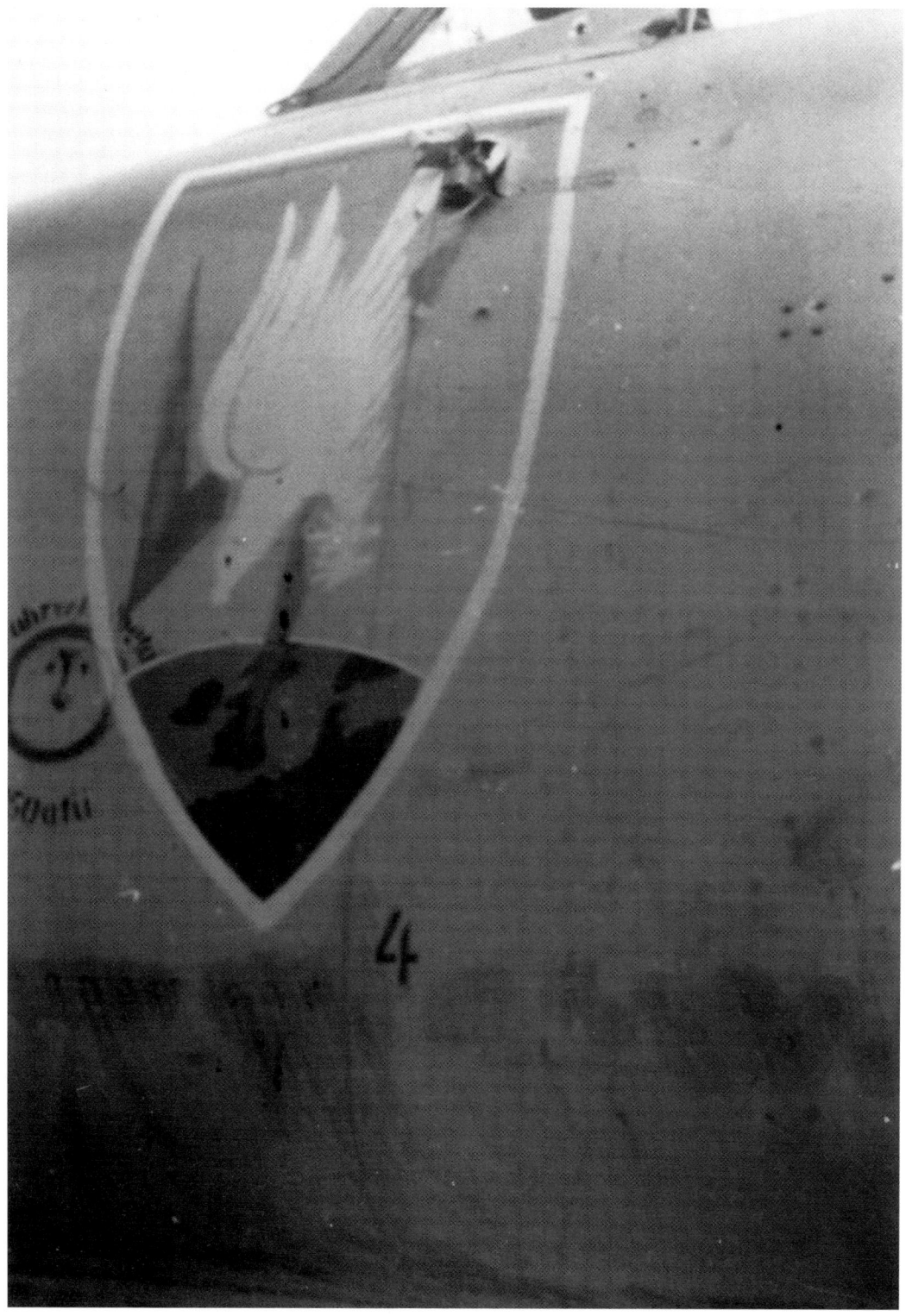

A close up of the 'England Blitz' badge on an unidentified Do 217 night fighter that was captured on the ground at Melsbroek, Belgium, in late 1944.

Do 217 M-11 of III./KG 100 on fire after crashing in the Gironde Estuary on 21 July 1944, while Mosquitos of 235 Squadron circle above. (via Andy Bird)

A Do 217 M-1 of 4.Nachtaufklärungsstaffel pictured flying over a wintry Russian landscape, 1944. (via Del Davis)

Personnel from 4.Nachtaufklärungsstaffel parading in front of a Do 217 M-1. (via Del Davis)

Loading BLC 50/A photoflash bombs onto a Do 217 M-1 K7+CM of 4.Nachtaufklärungsstaffel. (via Del Davis)

Do 217 K7+CM in flight. It is believed that this aircraft, Wk Nr 426343 K7+CM, was reported missing off Finland on 12 June 1944. (via Del Davis)

Groundcrew at work on a Do 217 M-1 of 4.Nachtaufklärungsstaffel. (via Del Davis)

Guarding a Do 217 M-1 of 4.Nachtaufklärungsstaffel. Note the distinctive spinners. (via Del Davis)

Refuelling a Do 217 M-1 of 4.Nachtaufklärungsstaffel. This unit was still operational at the end of the war. (via Del Davis)

A group of Do 217 Js of NJG 101 photographed over Hungary in 1944.

This Do 217 N-1 of 12./NJG 5, coded C9+SW, was photographed in 1944. The Wk Nr 509 can be seen on the nose.

Covered in camouflage nets, this Do 217 M-1 of 4./Ergänzungs Fernaufklärungsgruppe was photographed at Brieg in the autumn of 1944.

An aged Do 217 E-1 with the Wk Nr 094 on the nose of FFS(C)5. It is possible that this was Wk Nr 5094, which was lost in an accident at Usadel, Tollensee, on 7 February 1943. Leutnant Harald von Lack and the Bordmechaniker, Unteroffizier Walter Börner, were killed.

Do 217 E-4 Wk Nr 5369 of 3./Ergänzungs Zerstörergruppe Vaerløse. This aircraft suffered an accident at Gilze-Rijen, while with I./KG 2, on 7 August 1942.

Postscript

With the production of the Do 217 ending in October 1943, further development ceased. It had been intended to have a M-2 (torpedo), M-3 (dive bomber), M-4, M-8 (high-altitude bomber), M9 and M10 (guided missile carrier) versions, but these projects did not make it off the drawing board or were cancelled. However, forty Do 217 M-11s, which were designed especially for the Fritz X and Hs 293 missiles, were built and were used operationally in very limited numbers by KG 100 between June and August 1944.

Of the Do 217, one German pilot wrote: 'By July 1941, it was still not ready to be deployed to the Front. Many inadequacies were still to be discovered and it required constant changes, improvements and refitting. In all of the years of the war, we were hardly ever at full strength and escalating losses played a major role with this.'

At the end of the war, Do 217s were scattered around numerous airfields in Europe, all of which were soon scrapped. As a result, there are no complete examples in existence today.

Perhaps the final word on this aircraft should go to test pilot Captain Eric 'Winkle' Brown. Eric flew what he nicknamed 'The Flying Pregnant Pencil' after the war. He duly described the Do 217 M as 'no world beater', and concluded by saying: 'The Do 217 was a moderate aircraft which established an undistinguished but honourable record.'

Although trialled for torpedo dropping, the Do 217, such as that K-1 seen here from the torpedo trials unit at Gotenhafen, was not deemed suitable.

This Do 217 K-03, coded KE+JC, was used as a test bed for the DFS 228 high-altitude rocket-propelled aircraft.

Postscript 227

Do 217 PV1 Wk Nr 1229, coded BK+IR, was designed as a high-altitude bomber with a pressurised cockpit and powered by two DB 603N engines with a DB 605 booster.

Do 217 PV1 Wk Nr 1229 first flew on 6 June 1942. By 1944 further development work was cancelled in favour of other types.

The Do 317 V1 prototype VK+IY was powered by DB 603B engines and had triangular tail fins. Bigger and more spacious than the Do 217, it was a contender for the next Luftwaffe bomber, the so called Bomber B. It first flew in September 1943, but further development was halted.

At the end of the war, many destroyed aircraft were found on airfields used by the Luftwaffe. This shows the remains of U5+AM of 4./KG 2 at Gilze-Rijen.

In this further view of U5+AM of 4./KG 2 at Gilze-Rijen, the 'U5' is visible ahead of the cross, while 'AM' can be seen in white on the tail.

This Do 217 M-1 was found at Prague-Ruzyne airfield. The code is EK, which would probably mean it is from 2.Nachtaufklärungsstaffel, the full code being K7+EK.

This Do 217 M-1, coded K7+LH, is from 1.Nachtaufklärungsstaffel and was captured at Copenhagen. Many Nachtaufklärungsstaffel aircraft had spirals painted on their spinners.

Seen at Odense (Beldringe) in Denmark are three of six Do 217 M-1s captured there at the end of the war. The men and machines of 2.Nachtaufklärungsstaffel moved to Odense on 1 May 1945, where they were captured upon the German surrender. It has been stated that the middle aircraft is U5+HK Wk Nr 56527; the letters +HK are visible on the fuselage. This aircraft was exhibited at Farnborough in October and November 1945, being allocated the Air Ministry number 106.

This Do 217 M-1, coded K7+AH, of 1.Nachtaufklärungsstaffel was captured at Linz-Hörsching. Note the spirals on the spinners. It subsequently scrapped.

Just visible on the dump at Linz-Hörsching is the Do 217 M-1 that was coded K7+AH.

Do 217 M-1 Wk Nr 56158 is believed to have been one of the aircraft from 2.Nachtaufklärungsstaffel captured at Odense (Beldringe) in May 1945.

Do 217 M-1 Wk Nr 56158 was subsequently flown to Farnborough, given Air Ministry number 107 and displayed. However, by December 1945 it had been placed in storage, initially at RAF Brize Norton and then at RAF Sealand. It is believed to have been scrapped at RAF Bicester in 1956.

Where it all began – Do 217 E-1 Wk Nr 1037, coded TC+ZK, undergoing trials with Dornier. It was badly damaged in a crash-landing at Lechfeld, while being operated by II./KG 40, on 29 June 1941.

Dear Reader,

We hope you have enjoyed this book, but why not share your views on social media? You can also follow our pages to see more about our other products: facebook.com/penandswordbooks or follow us on X @penswordbooks

You can also view our products at www.pen-and-sword.co.uk (UK and ROW) or www.penandswordbooks.com (North America).

To keep up to date with our latest releases and online catalogues, please sign up to our newsletter at: www.pen-and-sword.co.uk/newsletter

If you would like a printed catalogue with our latest books, then please email: enquiries@pen-and-sword.co.uk or telephone: 01226 734555 (UK and ROW) or email: uspen-and-sword@casematepublishers.com or telephone: (610) 853-9131 (North America).

We respect your privacy and we will only use personal information to send you information about our products.

Thank you!